J. Andy Soesilo
Stephanie R. Wilson

Hazardous Waste Planning

CRC

LEWIS PUBLISHERS

Boca Raton New York London Tokyo

Library of Congress Cataloging-in-Publication Data

Soesilo, J. Andy.
Hazardous waste planning / by J. Andy Soesilo, Stephanie R. Wilson
p. cm.
Includes bibliographical references and index.
ISBN 0-87371-497-0
1. Hazardous waste—Management. 2.Hazardous waste sites.
I. Wilson, Stephanie R. II. Title.
TD1030.S64 1995
363.72′875—dc20 94-38097
CIP

Direct all inquiries to CRC Press, Inc., 2000 Corporate Blvd., N.W., Boca Raton, Florida 33431.

Lewis Publishers is an imprint of CRC Press

International Standard Book Number 0-87371-497-0
Library of Congress Card Number 94-38097
Printed in the United States of America 1 2 3 4 5 6 7 8 9 0
Printed on acid-free paper

Preface

This book describes the management of hazardous waste programs in the private and public sectors from the planning perspective. The discussion covers a variety of planning components including the statutory and regulatory framework, data and information management, and types of hazardous waste planning.

In general, the book deals with problems and issues confronting individuals in the hazardous waste management and planning profession. Most important, the book provides approaches to addressing these problems.

The book presents the fundamentals of hazardous waste planning. It systematically compiles various sources of documentation in the field of hazardous waste management and planning and blends them into an integrated planning text.

The book is intended for use by the industrial community and consulting firms engaged in the management of hazardous waste, state and local environmental agencies, environmental planning students, environmental groups, and anyone who wishes to learn more about the basics of hazardous waste planning. To satisfy this objective, the book was written in as detailed and comprehensive a manner as possible, but with simplicity as its main attribute.

Biography

J. Andy Soesilo, Ph.D. is currently the manager of the Hazardous Waste Section at the Arizona Department of Environmental Quality. He is responsible for the management of the state's RCRA Subtitle C program, which encompasses hazardous waste planning, data management, permitting, inspection, enforcement, and emergency response. He is also an adjunct professor at the Center for Environmental Studies at Arizona State University. Dr. Soesilo received his B.S. (1972) in civil engineering from Petra Christian University, his M.S. (1976) in urban planning from the Asian Institute of Technology, and his Ph.D. (1986) in geography from Arizona State University.

Stephanie R. Wilson is currently the manager of the Underground Storage Tank (UST) program at the Arizona Department of Environmental Quality (ADEQ). She oversees the UST prevention activities of inspections and compliance; the LUST corrective action efforts of remediation and enforcement; and the State Assurance Fund reimbursement claims. Previous jobs included Acting Assistant Director for the ADEQ waste programs and lead planner for the ADEQ Waste and Air Quality programs. Ms. Wilson received her B.S. in zoology (1970) and her M.S. in biology (1973) from Kent State University.

Acknowledgments

The authors would like to give thanks for reviewing parts of the manuscript to **Linda J. Allen, P.E.,** Pollution Prevention Technical Assistance Coordinator, and **Dennis C. Clayton,** Technical Programs Unit Manager, who are both with the Arizona Department of Environmental Quality. Our appreciation for manuscript review is also given to **Douglas Kukino,** Environmental Administrator for the City of Glendale, Arizona, and **Daniel Roe,** Executive Director of the Arizona Emergency Response Commission. Finally, special recognition is given to **Professor Martin J. Pasqualetti** of the Department of Geography and to **Professor David K. Pijawka** of the Department of Planning, both at Arizona State University, who provided insight and valuable comments to the structure of the text.

Contents

Part One — Introduction

Part Two — Legal Aspects

Part Three — Data Management

Part Four — Types of Hazardous Waste Planning

Part Five — Hazardous Waste Planning Trends

Part One — Introduction

Hazardous waste planning has evolved as a challenging profession amid a series of intense technical, social, economic, and political debates. The controversies have focused on the necessity of economic growth versus community reluctance to accept hazardous waste as the price of growth. The controversies indicate that hazardous waste problems do exist. Fortunately, like all traditional planning practices, hazardous waste planning is a problem-solving activity.

Part One of this book contains two chapters. Chapter 1 discusses in detail an array of hazardous waste planning problems to provide an understanding of the dilemma facing hazardous waste planners. Chapter 2 examines the elements of hazardous waste management planning. The review covers fundamental planning issues, hazardous waste management philosophy, hierarchy, objectives, and planning requirements. A brief discussion of hazardous waste planning typology is also provided.

1 Problems and Issues in Hazardous Waste Management

The purpose of this book is to provide the readers with an overview of the field of hazardous waste planning in the private sector and in governmental agencies. Throughout the text, the management of hazardous waste is discussed within the context of planning. In this book, hazardous waste planning and management are considered to be two closely related functions.

Like all traditional planning practices, hazardous waste planning is a problem-solving activity that identifies problems, finds and plans the best solution, and implements the plan. This simplified procedure does not ignore the fact that hazardous waste management is a highly complex process. It is a mix of science, engineering, and data management, and includes legal, administrative, societal, and policy decision-making considerations.

In order to understand the dilemma of the profession, this book starts with identifying the problems and issues in the field of hazardous waste management in the industrial community and in the government. Because this book is intended for both corporate and governmental agency planners, a separate discussion on the interaction between the two planning fields is provided at the end of this chapter. The magnitude of hazardous waste problems is discussed in the first part of this chapter.

THE MAGNITUDE OF THE HAZARDOUS WASTE PROBLEM

Generally, there are a number of ways to describe problems in the management of hazardous waste. The most-used approach in identifying hazardous waste problems is to trace the problems back to their historical, regulatory background. The history of waste management in the United States essentially parallels the country's progress in economic development. The U.S. began its economic surge with the industrial revolution in the latter part of the 19th century. Rapid growth continued through the early 20th century, fueled by the defense demands of World War II. After the war, the nation's revitalized industrial base shifted from armaments to the production of consumer goods.

The production of both military and consumer goods generated major quantities of municipal refuse and industrial waste. The ability and concern to manage waste in an environmentally protective manner were lacking. Garbage generated by Americans in their daily lives made its way into the environment, posing a serious threat to ecological systems and public health. This situation prompted Congress to pass the Solid Waste Disposal Act in 1965. The Act provided funding to state and local governments for the development of safe garbage disposal methods.

In the 1970s, environmental problems caused by improper management of hazardous waste generated by chemical and other industrial processes captured the public's attention and resulted in a heightened public concern of its potential dangers. Hazardous waste that is spilled or improperly disposed of has severely contaminated soils and surface- and groundwater, killing wildlife and aquatic organisms. It has also posed dangers to human health, ranging from minor headaches to terminal cancer.

In response to the increasing awareness of the serious problem stemming from improper disposal of hazardous waste, Congress amended the Solid Waste Disposal Act and enacted the Resource Conservation and Recovery Act of 1976 (RCRA), which established a *cradle-to-grave* system of hazardous waste management. As the name implies, the system provides a regulatory framework for the management of hazardous waste from the point of generation to the final disposal. The Act remains the foundation of current management of hazardous waste.

Managing hazardous waste in the 1990s represents a unique challenge for hazardous waste planners and managers. This is shown by the variety of problems related to the management of hazardous waste. RCRA defines hazardous waste as a solid waste — or combination of solid wastes — which, because of its amount, concentration, or physical or infectious characteristics, may pose a hazard to human health or the environment when improperly stored, treated, transported, or disposed of. Because of the complexity of the process, the identification and determination of hazardous waste is specifically discussed in Chapter 3.

Some of the major current problems and issues associated with hazardous waste management include:

1. the ever-increasing strictness of hazardous waste regulations and the increasing emphasis on enforcement and higher penalties,
2. the "Not in my backyard" (NIMBY) syndrome and other problems associated with hazardous waste facility siting,
3. the hazardous waste contamination from inactive hazardous waste sites,
4. the soaring costs of hazardous waste management,
5. the difficulty of exporting hazardous waste to other states,
6. the quality of waste generation data,
7. the amount of hazardous waste generated, and
8. maintaining progress amid economic development.

This listing is obviously not completely inclusive, but it provides planners with some current hazardous waste planning issues and with the magnitude of the problems that must be considered when doing their jobs. The following sections discuss each of the problems in more detail.

Stricter Hazardous Waste Regulations and Enforcement

RCRA was amended by the Hazardous and Solid Waste Amendments (HSWA) in 1984. The amendments contained four new provisions:

- businesses generating small amounts of hazardous waste were required to comply with certain provisions of the law;
- certain types of hazardous waste were prohibited from land disposal without prior treatment;
- source reduction and recycling were viewed as preferred management methods; and
- stricter standards and regulations were established for hazardous waste handlers.

In view of the above amendments, it is apparent that national hazardous waste management concerns caused standards and regulations applicable to hazardous waste to become more stringent. Additionally, governmental enforcement efforts also became more rigorous.

Changes in the enforcement program can be best recognized by examining the RCRA enforcement statistics reported in 1990 (Figure 1.1) by the U.S. Environmental Protection Agency (EPA).[1] The figure shows that during the 1980s, RCRA enforcement steadily increased from only 53 actions in 1980 to 1,940 in 1989. Prior to 1985, the actions covered administrative orders and judicial referrals enforced by the federal government. Since 1985, the enforcement efforts in these two areas have been undertaken by both the federal government and the states.

Administrative orders are issued by the EPA (or a state environmental agency) to hazardous waste handlers to address rule violations and to force compliance within a given period of time. Ignoring an administrative order may result in the issuance of a judicial referral in which the EPA or a state environmental agency refers the case to the court for further judicial process.

In 1989, a total of 1,642 administrative orders were issued along with 298 judicial referrals. Judicial referrals are either civil or criminal, depending upon the severity of the violations. Civil referrals may result in civil penalties, while criminal referrals may result in criminal penalties. Both types of penalties can be severe and may involve federal employees as well as the public sector (see Table 1.1).

The extent of civil and criminal penalties is prescribed in federal and/or state laws. State civil penalties may result in fines as high as $50,000 a day per violation (in New York, New Jersey, and Louisiana). State criminal penalties may result in imprisonment of 8 years and/or fines of $500,000 such as in Colorado, and 15 years and/or $250,000 in Texas.

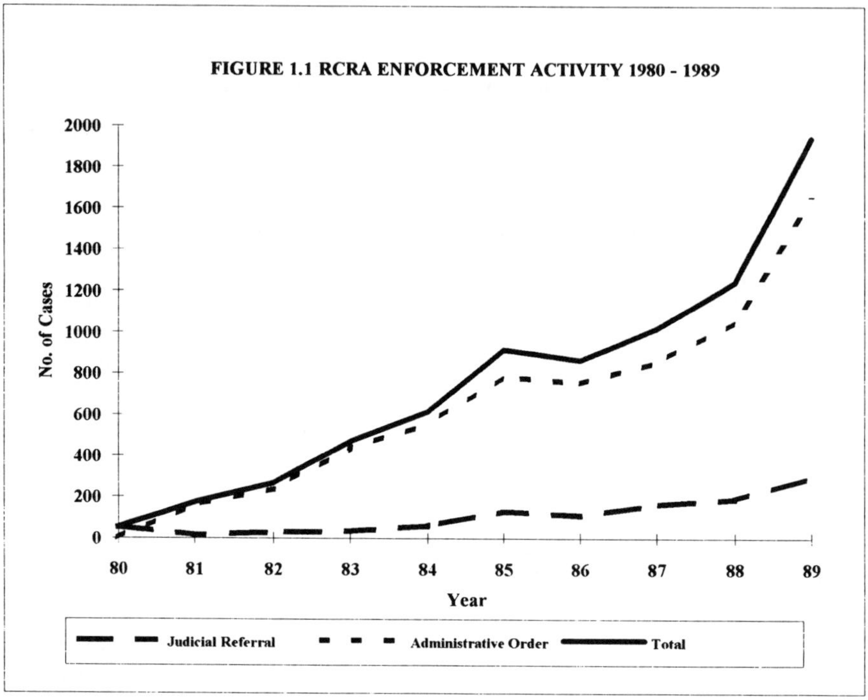

Figure 1.1 RCRA Enforcement Activity, 1980–1989. (From U.S. Environmental Protection Agency, *Enforcement Accomplishments Report,* U.S. EPA, Washington, D.C., 1990.)

The federal government's firm commitment to enforce RCRA laws is reflected by its willingness to pursue violations utilizing criminal enforcement authorities and by integrating criminal referral into the enforcement program. The states, however, have indicated concerns that imposing tougher enforcement in one state may place industries in that state at a competitive disadvantage vis-à-vis similar businesses in other states that do not have similar enforcement policies. Tougher enforcement may also dissuade businesses from locating in the state (U.S. GAO, 1991).[2]

The commitment to pursue tougher enforcement, combined with public awareness of waste-related issues, will certainly affect the way the private and public sectors manage hazardous waste now and in the future. The public's concern about hazardous waste goes beyond enforcement and includes the siting of hazardous waste facilities. Opposition to new facilities will also affect hazardous waste management decisions. The following section briefly describes the NIMBY syndrome.

Table 1.1 Examples of Enforcement Actions

Case #1. Six Million Dollar Penalty

On August 31, 1993, the U.S. District Court for the Northern District of Indiana fined Bethlehem Steel Corp. $6 million for "long-standing and willful violation" of RCRA at the company's plant in Burns Harbour, IN (Environment Reporter, 1993).[3] As of early 1994, the $6 million fine represents the largest monetary penalty ever imposed under RCRA.

Case #2. Imprisoned for Illegal Dumping

A case involving a Mississippi man in federal court of illegally dumping hazardous waste provides an example of a violation resulting in a criminal penalty. The man was sentenced to nearly three and a half years in prison without parole. This was the first sentence in Mississippi for spoiling the environment (Environment Reporter, 1990).[4] The indictment charged that the man was responsible for dumping sixteen 55-gallon drums of hazardous wastes into a creek in Jones County.

Case #3. Federal Employees Convicted

In a 1989 verdict by a federal jury in Maryland, three high-level civilian managers at the U.S. Army Aberdeen Proving Ground were convicted of illegally storing and disposing of hazardous waste at an aging chemical weapons research plant. Each person was sentenced to three years probation and was ordered to perform 1,000 hours of community service. Hazardous wastes were stored for years without a permit. Some of the chemicals were stored in rusty, leaking drums or in unheated buildings. It was the first time federal employees were convicted of hazardous waste violations under RCRA. The verdict sent a "shock wave" through the federal work force (Environment Reporter, 1990[5]; WTN, 1989[6]).

The NIMBY Syndrome and Hazardous Waste Facility Siting

The first Earth Day on April 22, 1970 sparked America's modern environmental movement. Particularly during the late 1980s, environmental groups actively participated in events across the country to demonstrate concern for the environment (Americans for the Environment, 1989).[7] Community resistance to the siting of hazardous waste management facilities represents a vivid example of the notorious "Not in my backyard" (NIMBY) response so familiar to planners struggling with the location of *locally unwanted land uses* (LULU) that are regionally necessary (Popper, 1985).[8]

The NIMBY response to the proposed location of a hazardous waste facility has spread so widely that Heiman (1990)[9] perceives it as a social phenomenon expanding beyond local resistance to a more trenchant "Not in *anybody's* backyard" syndrome. Heiman argues that no one can deny that community opponents to a hazardous waste facility have greatly strengthened their solidarity and their ability to thwart siting proposals.

A national survey prepared for the New York State Legislature revealed that, of the 81 applications for new facilities subject to federal and state regulations from 1980 to June 1987, only six applications across the entire nation resulted in new operating facilities. Moreover, of the six operating plants, four are hazardous waste treatment and transhipment centers — which are perceived as having a lower risk to public health and the environment than incineration or landfill operations. These four plants are located in Alabama, Florida, Rhode Island, and Texas. The remaining two new facilities are a landfill in Missouri that has been closed until an expansion approval is received and a hazardous waste incineration facility in Illinois (New York State Legislature, 1987).[10]

Community opposition to hazardous waste facilities is partly triggered by the risk associated with the management of hazardous waste, and improper handling and past mismanagement that created severe contamination at inactive sites discovered two decades ago.

HAZARDOUS WASTE CONTAMINATION OF INACTIVE SITES

In the 1970s, the American public started to realize that the hazardous waste problems caused by past mismanagement were not addressed by the existing environmental statutes. In that decade national attention was focused on the discovery of a large number of abandoned, leaking hazardous waste dumps that were contaminating the environment and threatening human health. These chemical dumps represented a series of "toxic time-bombs waiting to go off" and were scattered across the country from New Jersey to Delaware, Michigan, Iowa, Tennessee, Louisiana, California, Oregon, and Washington (Brown, 1980).[11]

The first major incident was **Love Canal** in Niagara Falls, NY, where people were evacuated from their homes after hazardous waste — buried by a chemical company for over 25 years — seeped to the surface and into basements. The canal was originally designed to transport water, and during the 1940s it was capped with clay and soil after being used for burial of hazardous waste. In the early 1950s a community was built around the canal. Over the years the canal leaked, and by 1978 the danger to the residents had become so great that President Carter was forced to declare the canal area a national disaster (Silverman, 1989).[12]

In 1988, a Love Canal habitability study was completed after 6 years of research. The $16 million study found that, after being cleaned up, two-thirds of the Love Canal neighborhood would be suitable for residential use, while the other third could be used for industrial or commercial purposes. Based on the study's findings, the Love Canal Area Revitalization Agency developed a master plan. Residential resettlement started in September 1990 (Environment Reporter, 1990[13]; Silverman, 1990[14]).

Times Beach, MO, illustrates another outstanding story of past hazardous waste mismanagement. In this case, oil that was contaminated with a hazardous waste called dioxin was used to pave local roads. The hazardous waste subsequently contaminated the soil and groundwater in the community (Vig and Kraft, 1983).[15]

Residents of Times Beach were relocated by the federal government in 1983 at a cost of $33.5 million. As of 1990, an estimated $52.5 million has been spent to clean up Times Beach. Syntex Agribusiness was named one of several potentially responsible parties. On June 20, 1990, the company agreed to continue the cleanup project, which was estimated to cost between $90 million and $110 million in addition to reimbursing the federal government $10 million of the estimated $86 million it had spent. The remediation plan included the incineration of dioxin-contaminated soil. The remediation is expected to be completed in 1997 (Superfund Report, 1991[16]; Suzanne DiPasquale, 1991[17]; Environment Reporter, 1990[18]).

The discovery of hazardous waste–contaminated sites resulted in an awareness that a disaster of this type and magnitude was outside existing environmental laws. A Congressional committee survey found that one-third of the 3,383 waste disposal sites that had been used since 1950 by the 53 largest U.S. chemical companies were not subject to federal regulations (U.S. EPA, 1988).[19]

In response to the need for the cleanup of hazardous waste in abandoned sites, the Comprehensive Environmental Response, Compensation, and Liability Act (CERCLA) — popularly known as *Superfund* — was enacted in 1980. The Act authorized a tax to be levied on the chemical industry. The Superfund program helped to clean hazardous waste sites by removing contaminated soils from sites and/or by remediating the contaminated sites by using various treatment technologies.

The Superfund program requires the EPA to identify the parties responsible for the contamination of a site. The potentially responsible parties (PRPs) include past and present owners or operators of the site, generators of hazardous waste found at the site, and transporters of hazardous waste to the site. Once the PRPs are identified, the EPA can negotiate a legal agreement or can unilaterally order them to take remedial-action responsibility of the site.

In 1989, a group of 533 PRPs in Michigan signed a consent decree in federal court for the cleanup of the contaminated Liquid Disposal Inc. site. It was the largest group of PRPs to sign a consent decree in the history of Superfund during the 1980s (WTN, 1989).[20] The consent decree contained the PRP's agreement to pay the EPA $24.4 million for the cleanup of the site. In return, the court dismissed the case.

In the same year, the U.S. District Court approved a settlement valued at more than $66 million with over 100 PRPs for the Operating Industries Inc. site in Monterey Park, CA. The settlement was one of the largest monetary settlements reached in the Superfund program in the 1980s (U.S. EPA, 1990).[21]

Yet, another consent decree signed in 1990 required a former steel company to pay at least $22 million for cleaning up contaminated sites south of Salt Lake City, UT. This case was the nation's largest settlement involving a bankrupt party held liable for a hazardous waste–contaminated site (WTN, 1990).[22]

These examples illustrate various Superfund enforcement actions by the federal government. The intensity of the Superfund program is reflected in the following EPA figures. From 1984 to 1989, the agency issued a total of 1015 CERCLA administrative orders, an average of 170 orders a year (U.S. EPA, 1990).[23] These figures do not include cleanup actions initiated by states. State actions to clean up abandoned sites are triggered when the federal government is slow in responding to local needs.

Cleaning up a contaminated site, as we can see from the preceding discussions, is a long and painful process. Most of the cleanup projects are very costly. For managing hazardous waste in a currently operating facility, the cost is expensive and increasing over time. This is one of the reasons why proper, cost effective planning and management of hazardous waste is necessary.

Soaring Costs of Hazardous Waste Management

Stricter environmental standards and regulations, tougher governmental enforcement programs, the liability potential for improperly managed hazardous waste, and the public rejection of hazardous waste facility siting proposals have all contributed to the high and rapidly growing cost of managing hazardous waste. This trend is expected to continue in the future.

Surveys of selected firms in the commercial hazardous waste management industry indicated that from 1983 to 1987, the average charge for most hazardous waste handling services rose by at least 69 percent, and some services increased by as much as 1940 percent over the same period (Forcella, 1989).[24] The surveys documented that landfill disposal costs that were between $25 and $60 per drum in 1983 increased to $64 to $186 per drum in 1987. Incineration of highly toxic liquids increased from a range of $1.50–$3.10 per gallon in 1983 to $2.36–$5.02 in 1987.

Costs have escalated in almost all disposal categories, but now many industries also face a prohibition on land disposal, which traditionally has been the least expensive disposal option. The land-ban regulations restrict numerous wastes from land disposal without prior treatment (U.S. EPA, 1989).[25] Treatment standards are set for each category of hazardous waste, generally as maximum concentration levels of hazardous constituents based on the performance of *best demonstrated available technology* (BDAT). Treatment standards are also expressed as *specific technologies* when data are inadequate to set concentration levels, or when it is determined that one method — such as incineration — is the best way to treat the waste. These land disposal restrictions will force industries to use more expensive treatment technologies for many types of hazardous wastes.

The land-ban regulations group hazardous wastes into five categories based on a scheduled compliance date. The first regulation, which took effect on November 7, 1986, controlled spent solvents and dioxin wastes. The second group is known as the *California List* wastes. The list includes liquid wastes containing certain metals, free cyanides, polychlorinated biphenyls (PCB), corrosives with a pH of less than or equal to 2.0, and liquid and nonliquid hazardous wastes containing specified halogenated organic compounds (HOC). The California List wastes became effective on July 8, 1987.

The remaining three groups are referred to as the *third* land-ban wastes. The *First Third* wastes consist of 121 hazardous wastes that are restricted from land disposal and 62 hazardous wastes that are banned from land disposal. The regulations took effect on August 8, 1988. The *Second Third* rules, which became effective on June 8, 1989, established treatment standards for at least 67 additional hazardous wastes. The *Third Third* rules impose restrictions on 349 listed hazardous wastes. The phased implementation of the Third Third regulations was completed by May 8, 1992. The land-ban rules, when totally implemented, affected most of the approximately 700 wastes currently regulated under RCRA.

Industry can also expect disposal costs to increase as (1) regulations are promulgated to encompass an expanding universe of hazardous waste; (2) more disposal restrictions are enacted; and (3) specific treatment technologies are prescribed for various wastes.

An April 1990 U.S. General Accounting Office report (U.S. GAO, 1990)[26] described the EPA's information on threats to groundwater posed by nonhazardous waste facilities and evaluated the EPA's plan to revise its standards for these facilities. Nonhazardous wastes are those that either do not meet the EPA's definition of hazardous waste or are exempted from hazardous waste regulations. Nonhazardous wastes include certain industrial and municipal wastes, construction/demolition debris, mining and ore-processing wastes, cement kiln dust, byproducts of fossil fuel combustion, and drilling fluids and other wastes from oil and gas production. The report concludes that some of the wastes are dangerous enough for the EPA to plan to regulate them as hazardous in the near future.

The Toxicity Characteristic Leaching Procedure (TCLP), which was promulgated on March 6, 1990, requires a more stringent laboratory procedure to determine whether a waste is hazardous. As a result of the promulgation of TCLP rules, more than 200 million tons of wastewater are expected to be classified as hazardous, and about 17,000 additional waste generators in 12 major industries will be affected by the new testing procedure. The largest cost burden is expected to fall on five industries (WTN, 1990)[27]:

- petroleum refining,
- pulp and paper,
- synthetic fibers,
- wholesale petroleum marketing, and
- organic chemicals.

Confronted with the growing treatment and disposal costs and stricter environmental regulations, certain industries have reacted in a positive way. Some companies have already made decisions to integrate environmental considerations into their industrial operations, so they have designed strategies to minimize hazardous waste generation and to reduce potential liabilities.

Unfortunately, there are instances where industry continues to endanger public health and the environment by falsifying certifications of meeting a treatment standard or by illegally dumping hazardous waste despite the intensive enforcement actions from the regulatory agencies (Morgulec and Woitte, 1991).[28] This is partly due to the disproportionate number of hazardous waste handlers vis-à-vis the number of enforcement personnel.

Illegal dumping and improper handling of hazardous waste pose serious environmental problems, yet they depict only a part of the issues in the management of hazardous waste. From a public perspective, the balance between the amount of waste generated and the availability of hazardous waste facility capacity to manage the generated waste in the state requires special attention. The comparison between hazardous waste generation and treatment and disposal capacity demonstrates a need for additional state facilities to accommodate demand and/or an intensive waste minimization program.

Comparing the in-state demand for treatment and disposal capacities with its availability identifies whether a given state is an importer or exporter of hazardous waste. States with either no permitted treatment/disposal capacities or with inadequate capacities will have to export at least some hazardous waste.

Difficulty in Shipping Hazardous Waste to Other States

The export of hazardous waste has become a major issue between the states with hazardous waste treatment and disposal capacity and states without such capacity. Environmental Information Ltd. (1993)[29] reports that 37 states or territories were net exporters in 1989; 15 were net importers. States that imported the highest annual tonnage were Indiana (181,171 tons), Alabama (176,158 tons), and South Carolina (113,117 tons). The biggest exporters were California (176,342 tons) and Texas (108,820 tons).

Most imports and exports, according to the EI Ltd. study, occur between states that are near to each other. The complexity of the interchange is shown by the fact that on average, each state ships to 20 other states and receives hazardous waste from 20 states. The diversity of destinations represents the strongest argument for continued interstate waste movement, because generators are looking for different commercial services that provide environmentally protective hazardous waste treatment and disposal in a cost-effective manner.

Since the late 1980s, some states have tried to stop or limit the importation of hazardous waste through a number of regulatory actions. In 1989 Alabama enacted a controversial law that restricted the shipment of hazardous waste to facilities in that state. The law prohibited facilities in Alabama from accepting hazardous wastes from states that (1) did not allow disposal of such wastes within their borders, and (2) either lacked disposal facilities of their own or lacked a cooperative disposal agreement with Alabama (Cobb, 1990).[30] In a similar effort, South Carolina signed into law provisions limiting the amount of hazardous waste that can be imported into the state for disposal (Environment Reporter, 1990).[31]

As a consequence of more restrictive laws regarding hazardous waste transportation, costs for the existing permitted disposal sites have increased considerably. Alabama — home of Emelle, the 2.1 million-cubic-yard hazardous waste landfill owned by Chemical Waste Management, Inc. — has quintupled its waste import fee and has established a ceiling for the amount of waste that may be imported. The 1990 law increased the fee per ton a transporter must pay when disposing of hazardous waste in Alabama from $22 to $112. Alabama's action may prompt similar import fee charges in other states (Superfund Report, 1990).[32]

The 1993 Environmental Information report[33] indicates that there were 20,233 reporting generators in 1989. Most of these generators do not have on-site hazardous waste treatment or disposal capabilities. Statistically, over 95 percent of generators are completely dependent on off-site waste management facilities to handle their waste. EI Ltd. argues that in reality, virtually all generators rely on commercial hazardous waste management services for at least a portion of their needs; therefore, the imposition of fees by states would adversely affect the industrial community.

Interstate transportation of hazardous waste is costly. According to a 1988 survey by the Chemical Waste Transporters Institute, a total of 31 states assess some kind of transport fee. A company moving a load of hazardous waste from New York to California using Interstate 80 would have to pay approximately $2,500 in state permit fees to seven of the ten states on the route. A similar trip from Maine to Florida would require permits from nine of the ten states passed through and would also cost about $2,500. These costs do not include bonding requirements or state sales tax wherever applicable (Hilton, 1989).[34]

Although many states have enacted legislation to limit the importation of hazardous waste, the constitutionality of these laws is in question, and court action has already taken place. For example, a federal judge in January 1990 ruled that Alabama is not violating the U.S. Constitution by refusing to accept shipments of hazardous waste from states that do not have treatment and disposal facilities of their own or cooperative disposal agreements with Alabama. On August 8, 1990, a U.S. Court of Appeals overturned the earlier decision by ruling that the waste import barriers erected by Alabama are unconstitutional (Environment Report, 1991[35]; National Environmental Enforcement Journal, 1991[36]; Superfund Report, 1990[37]). The constitutionality of Alabama's hazardous waste import tax and South Carolina's hazardous waste import limitation law are still being debated.

Under the Commerce Clause of the U.S. Constitution, states are prohibited from regulating interstate commerce in all but a few limited circumstances. Becker (1989)[38] provides examples of limited state and local actions that have survived judicial scrutiny under the Commerce Clause. The examples are (1) restrictions on the use of private or public disposal facilities in response to a compelling, immediate health and welfare threat; (2) enactment of policies that limit the size and number of disposal facilities in a state to a level commensurate with state needs; and (3) imposition of disposal fees that are reasonably related to the cost to the state of accommodating disposal. Alabama and South Carolina cases may be judged unlawful under the Commerce Clause, but they have sent a clear message that each state must be prepared to manage the hazardous waste generated within its borders.

To properly manage hazardous waste, the waste management information and waste generation data must be accurate and reliable. In planning, the quality and reliability of data represents a significant planning requirement. Unfortunately, the quality of hazardous waste data is questionable and is still an issue. The following discussion on data quality is particularly focused on statewide data and therefore is applicable to governmental planning; however, corporate planners may be interested in this issue as well.

Quality of Waste Generation Data

There are major problems in comparing the demand for and supply of treatment and disposal facility capacities. The U.S. General Accounting Office (1987),[39] after completing a comprehensive national survey, concluded that it is difficult to accurately estimate the amount of hazardous waste produced — either nationally or at the state level — because of uncertainties in waste generation data. The study refers to methodological issues as the main reason for data inaccuracy. These include the different ways in which hazardous waste is defined and variations in estimating quantity.

In more recent studies, the General Accounting Office (1991[40] and 1990[41]) emphasized the need for improving hazardous waste data management. Under current law, the responsibility for data collection, as well as for other hazardous waste regulatory activities, are shared by federal and state governments. The variations in data management procedures among states diminish the quality of the national hazardous waste information.

RCRA encourages states to assume the responsibility for program implementation and enforcement from the federal government, because states are closer to and more familiar with local concerns and conditions.[42] Therefore, the states are in a better position to administer the programs and to respond to local needs. The Act was designed by Congress to have the entire hazardous waste program administered by states, and only minimal oversight would be provided by the federal government (U.S. EPA, 1990).[43]

An authorized state may then delegate its RCRA hazardous waste management authority to appropriate levels of local governments (city or county).[44] With the current trend to delegate RCRA authority from the state to local governments, a similar issue of nonuniform data management practices across intrastate regulatory agencies can be expected.

Although there are inaccuracies in hazardous waste information, the U.S. GAO estimates that the amount of hazardous waste generated over the years has been steadily increasing. The concern with this trend is very straightforward; i.e., the more waste generated, the riskier it is for the community. The next section discusses some of the aspects of hazardous waste generation.

Amount of Hazardous Waste Generated

An analysis based on the first Superfund Amendments and Reauthorization Act (SARA) — mandated Capacity Assurance Plan (CAP), submitted in 1989 by individual states, showed that the total amount of RCRA hazardous waste generated in the United States in 1987 was 228 million tons (Figure 1.2). The total quantity generated in 1989 was estimated to be 229 million tons, and the projected amounts for 1995 and 2005 were 235 million tons and 260 million tons, respectively.

In the CAP documents, some states provided data on successful accomplishments in their waste minimization programs. These states include, among others, Georgia, Illinois, Michigan, Missouri, New Jersey, New York, Ohio, Pennsylvania, and Virginia. Even with the waste minimization programs factored into the projections by some states, the projected figures based on the CAP methodology indicate a trend of increasing hazardous waste generation.

SARA requires each state to have an approved CAP to maintain eligibility for the receipt of Superfund monies to clean up contaminated sites. A CAP is

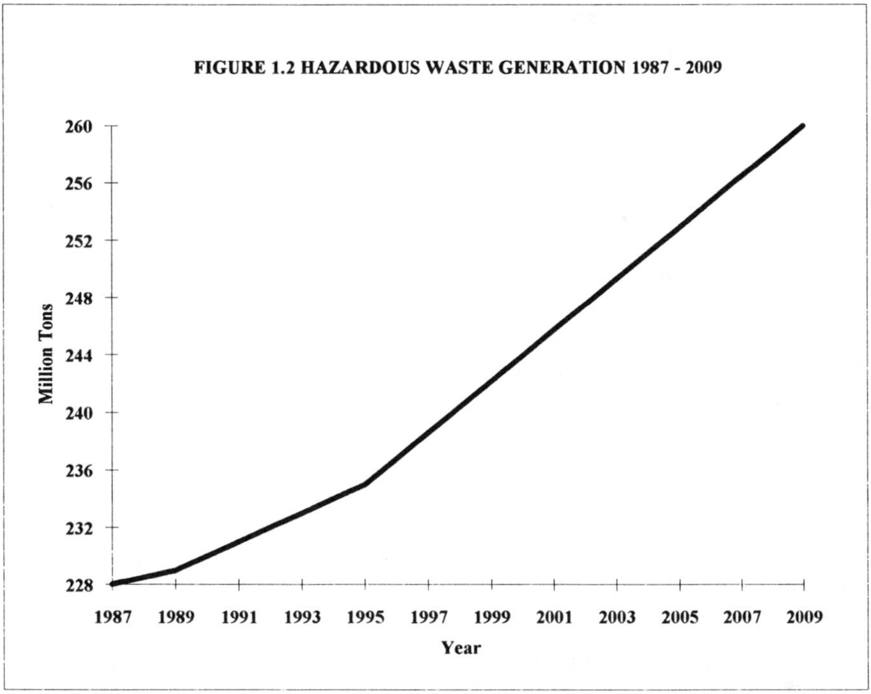

Figure 1.2 Hazardous Waste Generation, 1987–2009. (From 1989 CAP documents submitted by states.)

developed using data from the hazardous waste handler's biennial reports, which provide the best estimate of the amount of RCRA hazardous waste generated in the state.

The quantity of waste generated is a concern in the hazardous waste management field, because the amount of hazardous waste generated is related to both the level of risk to public health and safety and the adverse impacts to the environment. The risk is accentuated by improper management of hazardous waste, especially illegal dumping and transportation accidents involving hazardous waste.

The environmental and health risks associated with the generation and management of hazardous waste are a reality that every modern society must face. Although there are certain things that are under our control, a certain degree of risk exists as long as we continue to produce manufactured products. In fact, generating hazardous waste is part of our nation's economic activities. The central question to planners is to find ways to eliminate the risk and to maintain a proper balance between waste generation and progress in economic development.

Maintaining Progress in Economic Development

The two major areas of focus in hazardous waste planning and management are (1) reducing the generation rate and toxicity level of hazardous waste, and (2) efficiently managing the waste that has already been generated. Decreasing the waste generation rate and the toxicity level will reduce the risk to public health, to the environment, and to those people handling the disposal of the waste. Efficient waste management requires the safe handling of the waste at the least cost possible.

The goals of hazardous waste planning and management are to define the proper balance between protecting the environment and promoting economic development. These often-conflicting goals are obviously not easy to achieve nor always achievable. The economic development goal has both a positive and a negative relationship to environmental protection. Similarly, maximizing environmental protection may stimulate or hinder economic growth.

In recent years, the increasing size of hazardous waste problems, as shown in the previous sections, and the concern over economic growth have culminated to a point where better planning and management of hazardous waste is publicly demanded. Hazardous waste management professionals will undoubtedly play a more significant role in the future. Planners can contribute to this challenging field by actively participating in the environmental protection activities of corporations and governmental agencies.

HAZARDOUS WASTE PLANNING IN THE PUBLIC AND PRIVATE SECTORS

Protecting the environment and public health from hazardous waste contamination demands a collaborative effort by the public and private sectors. The public sector's responsibility is to develop and maintain a comprehensive hazardous waste management program. This can be achieved by first, establishing a regulatory framework based on the cradle-to-grave philosophy and through the implementation of various programs in the areas of permits, inspections, and enforcement. The role of hazardous waste planning for the public sector is to develop, implement, and enforce the effective implementation of the cradle-to-grave philosophy. This role, as later described in Chapters 4 and 11, is shared by the federal, state, and local governments.

The second part of a comprehensive hazardous waste management plan involves ensuring that existing local hazardous waste facilities have the capacity to handle the hazardous waste generated locally. This can be accomplished partly through a construction of new hazardous waste treatment and/or disposal facilities in the locality, and partly through governmental encouragement and assistance to industries to reduce hazardous waste generation as part of an industrial waste management strategy.

From an industrial viewpoint, a hazardous waste management program traditionally has been a compliance program that meets the hazardous waste regulatory requirements. An effective hazardous waste program must comply with the hazardous waste regulations at a "reasonable" cost to the company (Lindgren, 1989).[45] The cost may be significant due to the extensive nature and complexity of the current regulations.

As of July 1, 1990, the hazardous waste rules codified in the Code of Federal Regulations (CFR) parts 260 through 272 occupy 510 pages of the CFR, which is equal to approximately 700 pages of single-spaced-typed legal paper. The large number of rules in the CFR clearly indicates the extensive nature of hazardous waste regulations. They also are very cumbersome and hard to understand, especially for small companies.

Hazardous waste regulations are intricate and are continuously changing. Because the rules are being revised constantly, it is important that individuals and organizations be prepared to anticipate the changes. For example, the "third third" wastes' final rule, which was promulgated on June 1, 1990, contains approximately 400 pages of text and figures and represents one of the most massive RCRA rule changes (Environment Reporter, 1990).[46] The rules regulate the disposal requirements of over 344 hazardous wastes (Lowrance, 1990).[47]

In the 1990s the pace of environmental enforcement is expected to continue to increase, and governmental agencies will not hesitate to advocate holding corporate officers — rather than corporations — liable for environmental pollution. Although there are some fundamental principles that define the limits of personal civil or criminal liability of corporate officials, threats of personal fines and terms of imprisonment are now believed to be far more powerful disincentives to environmental noncompliance than corporate fines (Seymour, 1989).[48]

In anticipation of current and future environmental challenges, the successful corporations in the 1990s will be those that manage the company for environmental protection (McGregor, 1989).[49] According to McGregor, these firms do more than is required by laws and regulations. They anticipate the legitimate concerns of regulatory agencies and the public, and they subsequently address the concerns before such actions are mandated. In this way, they stay abreast or ahead of new government programs and policies, away from lawsuits, and ahead of the competition. By implementing environmentally protective programs before the law requires them and by designing the programs into projects, these forward-looking companies force the competition to catch up.

World Waste (1990)[50] has identified the 1990s as the decade of the waste management plan. Regulations, personal and financial liability, public pressure, and the cost of hazardous waste treatment and disposal will continue to drive industry toward better hazardous waste management practices. Corporate hazardous waste planning will be integrated into the overall corporate strategy.

Through the plan, the status and development of governmental hazardous waste planning will be continuously reviewed in order to stay ahead of the competition. Companies that understand the essence, nature, and scope of governmental hazardous waste planning have a competitive advantage, because these companies can use that knowledge to make effective corporate decisions in the area of hazardous waste management.

On the other hand, governmental agency staff should be more aware of the content and context of corporate hazardous waste planning. Such understanding is critical, because it can be used to assess the effectiveness of current governmental planning and, if necessary, to make modifications to the plan.

The complementary nature of the public and private hazardous waste planning professions suggests that, regardless of the place of a planner, there are basic planning components that every hazardous waste planner should know. The components include the hazardous waste legal background, the management of hazardous waste data, and the planning process itself.

2 The Scope of Hazardous Waste Planning

Traditionally, planning is viewed as an important part of management. McDowell (1986)[51] argues that not only is planning considered part of management, it is an integral part of the whole policy-making and implementation process. This approach necessitates that planners also provide information to decision-makers and to the general public, respond to regulations, and support the image or position of the organization (DeBettencourt et al., 1982).[52]

Hazardous waste planning is a function of hazardous waste management. However, planning and management in the domain of a hazardous waste business is so intricately integrated, it is sometimes difficult to differentiate between the two. This book uses the term "hazardous waste planning" interchangeably with "hazardous waste management." The central theme of this book, however, addresses the management of hazardous waste within the context of planning.

Like all planning, hazardous waste planning is a control of future action and a prescriptive, problem-solving activity (LeBreton and Henning, 1961).[53] As a problem-solving activity, planning starts with identification of what constitutes a planning problem. This planning process continues until a desired plan is formulated and implemented.

Depending on the nature of planning, different programs have different elements in their planning processes. Buskirk's (1986)[54] example of these elements is provided below. They include:

1. Identify problems and specify objectives.
2. Design, study, and compile an inventory of conditions and resources.
3. Analyze data and formulate plans and policies.
4. Evaluate alternative plans and policies, and select the best alternative.
5. Implement the plan or policy, and continuously monitor the implementation.

Because goal formulation and identification of objectives represent an initial stage of the planning process, this chapter will focus on this important topic. The three goals of hazardous waste management according to the Resource Conservation and Recovery Act (RCRA) are:

1. protecting the environment,
2. improving public health, and
3. conserving energy.

In order to achieve these goals effectively, what is needed is a strategy for implementing the hazardous waste management program. Once the strategy is given, hazardous waste planners are then required to formulate program objectives by integrating the goal and the strategy into the planning process. Before we go further with the initial step of the planning process, we must understand what constitutes the basic issue in the management of hazardous waste.

A FUNDAMENTAL HAZARDOUS WASTE MANAGEMENT ISSUE

The previous chapter discussed some of the problems and issues related to the management of hazardous waste. One of them is the general fear about the existence of hazardous waste in our society. In its concern about environmental risks caused by various types of hazardous waste, the public at large has often overlooked the basic principle of thermodynamics. The second law of thermodynamics states that, in converting energy to work, a certain amount of waste energy must be discharged into the environment.

Applying this law to hazardous waste suggests that the transformation process of raw material into final product involves the creation of waste, and in some cases, hazardous waste (Thibodeaux, 1990).[55] Looking at the transformation of raw material into products within the context of broader environmental management, Rowe et al. (1976)[56] consider environmental planning and management as the initiation and operation of activities to manage the acquisition, transformation, distribution, and disposal of resources in a manner capable of sustaining human activities, with a minimum disruption of physical, ecological, and social processes. In order to make an economic activity feasible, Rowe et al. suggest that the activity must be conducted at minimum environmental, monetary, and social costs. In this context, the goals of environmental management planning can be formulated to prevent adverse effects of economic activities with minimum monetary and environmental costs, in a just and equitable manner (Baldwin, 1985).[57]

Theoretically, Baldwin portrays the economic system as the transformation of material and energy resources into products and waste, and he postulates that the transformation occurs in a one-way flow. This portrayal is based on the *material balance model,* which equates resources as equal to products and services plus residuals. This is analogous to the law of conservation of mass and energy, which states that mass or energy *in* is equal to mass or energy *stored* plus mass or energy *out.*

Because the generated residuals eventually reach the biophysical environment and the affected environment may threaten the availability of resources, the material balance model provides an accurate picture of how an intense utilization of resources must be combined with careful residuals management. The model also explains the rationale of Baldwin's grouping of environmental planning into (1) natural resources planning and (2) residual planning.

Baldwin argues that residuals management not only deals with proper handling of residuals, but also considers any effort to reduce the generation of residuals early in the production design. He classifies residual planning as it relates to (1) water quality, (2) air quality, (3) solid waste, and (4) hazardous waste management planning.

Because most manufactured products generate hazardous waste, hazardous waste generation is consequently an integral part of our society's economic activities. The issue is not whether we are able to eliminate all hazardous waste, but perhaps more realistically, how well we manage that portion that cannot be eliminated.

Proper management means more than careful waste treatment or disposal. It includes the full range of management alternatives for each type of hazardous waste. The alternatives include, but are not limited to, the following (Daley, 1989[58]; Korte, 1988[59]):

1. reducing the quantity of waste generated at the source of generation,
2. treating the generated waste so it is no longer hazardous or is less hazardous,
3. in-house reusing of the waste in order to obtain the maximum use of process chemicals,
4. reprocessing the waste for reuse,
5. transferring the waste to another plant that may use the waste as raw material, and
6. separating hazardous waste from nonhazardous waste at the source of waste generation.

Choosing an appropriate option among a variety of unique hazardous waste management alternatives obviously represents a real challenge for planners. The selection is complex because it must consider technical, regulatory, and economic aspects. It is complex also because it must keep pace with the national hazardous waste management philosophy and integrate the philosophy into the planning program. The following discussions focus on the changes of the national philosophy of hazardous waste management.

A SHIFT IN HAZARDOUS WASTE MANAGEMENT PHILOSOPHY

Prior to the 1990s, the hazardous waste management philosophy focused primarily on improving technologies to manage hazardous wastes after they are generated. This philosophy is often referred to as the *end-of-pipe* approach. Among the currently available technologies are new equipment and sensors to accurately monitor pollutant levels. Advances in biotechnology have improved the effectiveness of treating hazardous wastes (Maltezou et al., 1989).[60] Future technological improvements that provide more efficient and "cleaner" processes will make significant contributions to the safe management of hazardous wastes. Cleaner and more efficient technologies offer effective solutions to detoxification and isolation of existing hazardous wastes from the environment (Tedder and Pohland, 1990).[61]

End-of-pipe treatment technologies generally require the addition of substantial amounts of chemicals and energy to remove the hazardous components in a waste stream (Campbell and Glenn, 1982).[62] According to Campbell and Glenn, end-of-pipe technologies are open-ended systems that use resources once and discard the waste products to an ultimate disposal facility.

Recently, there has been a shift in the national waste management philosophy toward preventing the generation of hazardous waste. This philosophy is referred to as the *source reduction* approach. The departure from the traditional end-of-pipe approach to this new philosophy is in response to more stringent regulations and the anticipation of potential future liabilities from improper hazardous waste management practices (Metry, 1989).[63] Current hazardous waste management involves more than mere regulatory compliance, as risk reduction and waste minimization measures are becoming important elements in such a program (Lindgren, 1989).[64]

Industries are now seriously considering not producing hazardous waste in the first place. If the production process cannot be modified to eliminate hazardous waste, many companies give the waste a "second life" by reusing it or recovering valuable resources from the waste. In preventing the generation of hazardous waste, a cross-media approach is necessary — i.e., to actually reduce waste material and not just transfer it from one medium to another — such as reducing hazardous waste but increasing air pollution (Hunt and Schecter, 1989).[65]

By integrating this cross-media approach into the source reduction and waste minimization program, we actually have come to the concept of pollution prevention. Detailed discussions on waste minimization and pollution preventions program will be provided in Chapter 7.

Some firms, mostly large corporations, adopted the source reduction concept long before the shift in approach occurred nationally. For these companies, the source reduction philosophy has been adopted as part of the company's culture. In 1972, for example, Dow Chemical U.S.A. introduced comprehensive Global Pollution Control guidelines, and in 1978 it started the Waste Reduction Always Pays (WRAP) program to reduce the company's hazardous waste generation (Martin, 1989).[66]

Another example is the 3M Company. The company's Pollution Prevention Pays (3P) program has been implemented since 1975. During the first year alone, the company prevented the generation of 397,000 tons of waste, 1 billion gallons of wastewater, 15,300 tons of water pollution, and 112,000 tons of air pollution. In the first 14 years of the program, the pollution prevented has resulted in a savings of $426 million for just the U.S. operations (Zosel, 1990).[67]

These companies illustrate success stories of firms that have adopted source reduction techniques. Many others are still developing better methods to reduce the generation of hazardous waste. Yet, some manufacturing firms are still not strongly motivated to implement source reduction technology. These companies are slow to develop source reduction programs because they are unfamiliar with available reduction technologies, reluctant to adopt something new, or unwilling to spend the time needed to carefully examine the program (Hansen, 1989[68]; Shorthouse, 1990[69]).

The implementation barriers for source reduction continue to persist. It is a challenge to hazardous waste planners to diligently plan a program that is able to overcome these obstacles. Conceptually, the source reduction approach is operative within a larger system. This larger system is called the *waste management hierarchy,* which describes the intent of the hazardous waste management program and the direction in which it should proceed.

HAZARDOUS WASTE MANAGEMENT HIERARCHY

The EPA in August 1976 stated that the desired approach to managing hazardous waste should be based on a hierarchy with source reduction as the first priority. This includes reducing the volume and toxicity of hazardous waste at the source through changes in the production processes. Next in the hierarchy is recycling, either through reusing the waste or recovering resources from the waste. This is followed by waste treatment, including incineration. Land disposal is considered the fourth and least desirable option because of the continuing threat of contaminant releases (U.S. EPA, 1988).[70]

The hierarchy provides an ordering of hazardous waste management technologies according to their possible environmental consequences. Source reduction is the most desirable option; but when source reduction options have been exhausted, recycling of the remaining waste should be the next option. Similarly, when recycling opportunities have been exhausted, then treatment methods should be explored before considering land disposal. In this context, it is not the order of the options that is often disputed, but the criteria by which options are judged to be exhausted (Wolf, 1989).[71] Theodore and McGuinn (1992)[72] argue that in reality, applying the hierarchy will not eliminate all wastes from all production processes; rather, it offers a more cost-effective means of minimizing the generation of waste. They observe that the four alternatives are required in a total systems approach to waste management.

Despite this stated preference, the EPA's efforts to promote source reduction have been limited. From 1976 to 1986, the EPA concentrated its activities on developing the hazardous waste regulatory program over the generation, transportation, storage, treatment, and disposal of hazardous waste, with a belief that the program would provide sufficient incentive for source reduction efforts (U.S. GAO, 1988).[73]

In its October 1986 report to Congress, the EPA pointed out that in managing hazardous waste, aggressive action in favor of preventing the generation of hazardous waste was needed, and that prevention was the only way to eliminate environmental risk (U.S. EPA, 1986).[74] The EPA again underlined — in a document published subsequently — that the biggest challenge in the 1990s in the area of hazardous waste management is the reduction of hazardous waste generation. The document indicates that hazardous waste management activities should be directed at investment in recycling programs, more efficient production processes, substitution of less harmful products, and the reduction of unnecessary wastes (U.S. EPA, 1988).[75]

Historically, the concept of waste reduction was first defined as waste minimization. It was introduced when Congress specifically stated in the Hazardous and Solid Waste Amendments of 1984 to the RCRA that it is the national policy of the U.S. that, wherever feasible, the generation of hazardous waste is to be reduced or eliminated as expeditiously as possible.

Later, the Pollution Prevention Act of 1990 established the hazardous waste management hierarchy as national policy. It declared that pollution should be prevented or reduced at the source wherever feasible, while pollution that cannot be prevented should be recycled in an environmentally safe manner. In the absence of feasible prevention or recycling opportunities, pollution should be treated. Disposal or other release into the environment should be used as a last resort.

The national hazardous waste management hierarchy represents a governmental preference of available hazardous waste management options. The hierarchy is a strategic instrument to achieve the environmental management goals of improving public health, protecting the environment, and conserving natural resources. In order to accomplish these goals, it is necessary to translate the hierarchy into a set of management objectives.

HAZARDOUS WASTE MANAGEMENT OBJECTIVES

When the national hierarchy was set up, the purpose was to overcome a number of impediments to the management of hazardous waste. The barriers to a hazardous waste management program include the lack of space for disposal facilities, the fear of the environmental and health risks associated with hazardous waste, and the limitations of hazardous waste technology and industry's monetary resources (Tanaka, 1990).[76]

The strategy has definitely stimulated the industrial sector to consider the hierarchy in its hazardous waste management practice. Although the hierarchy represents a preferred solution to hazardous waste management problems, the industrial community must consider other factors such as cost effectiveness of an alternative technology, organizational culture, environmental impacts of the technology, social acceptability of the technology, and the existing hazardous waste regulatory framework when adopting the hierarchy (Maltezou et al., 1989[77]; Wentz, 1989[78]).

In practice, a company must plan its hazardous waste management activities to meet the company's objectives and the government's regulatory requirement. A good plan is designed within the framework of the national hierarchy. The government can assist the private sector by providing technical assistance, regulatory interpretation, and information on record-keeping and reporting requirements.

The company's objectives may include minimizing total environmental compliance costs, minimizing corporate short- and long-term liability from hazardous waste, avoiding environmental damage from facility operations, fulfilling corporate responsibilities, or maintaining a good corporate image (Sutter, 1989[79]; Lindgren, 1989[80]).

All businesses, whether they provide a product or service, have one single purpose: to give customers something that is superior and delivered in a better way (Fox, 1990).[81] If business decisions are made on the basis of profit first rather than on the basis of customer satisfaction, the business will ultimately fail. Fox argues that profit is important to every business and that none can survive without it, but the reward of profit is the measure of how well a business is satisfying its customers and, as a result, its stockholders and employees. Profit is the consequence of providing superior goods and services. Protecting the environment should be viewed as a newly added customer requirement on the list of satisfactions. The requirement does not replace other satisfactions, but it does add one more important dimension which previously received less than adequate attention from businesses. Fox emphasizes that the survival of every business depends on how well this new need, along with satisfaction of other needs, is met.

Sell (1992)[82] reveals that managing hazardous waste will create some undesirable effects as well as beneficial effects. For example, it is going to cost increasingly more in the future to further improve our environment; yet, when we look at the progress that has already been made and at the tasks still facing us, it seems worthwhile to continue — with logic, knowledge, and good sense.

Knowledge that is required for planners to conduct proper hazardous waste planning includes, among other things, knowledge of hazardous waste statutes, regulations, and data requirements. These planning requirements are addressed in the following paragraphs.

HAZARDOUS WASTE PLANNING REQUIREMENTS

Toner (1988)[83] argues that the heart and soul of planning for most planners is in the regulations that implement their plans. In their profession, planners must continue to develop their planning skills and regulatory knowledge to keep pace with the intricate institutional setting and the rapidly expanding rules and regulations. This means that hazardous waste planners should have a good understanding of the legal framework of hazardous waste management. This significant aspect of planning is described in detail in Chapters 3 and 4. The legal background is discussed in terms of the laws, rules, and procedures related to the handling of hazardous waste.

While knowledge of rules and regulations is vital, hazardous waste data management represents another essential requirement for planners. As Slater (1984)[84] observes, many planning activities involve intensive regulatory analysis as well as advanced data management. The management of hazardous waste data and approaches to analyzing them are described in Chapters 5 and 6.

TYPES OF HAZARDOUS WASTE PLANNING

Depending on the specific planning objectives, hazardous waste planning creates plans, policies, or programs that can be classified into four major types of planning. They are:

1. hazardous waste management planning,
2. site correction and remediation,
3. emergency response and hazard management, and
4. citizen participation planning.

Hazardous waste management planning is an activity that supports the goal of hazardous waste management. The goal is to ensure the proper management of hazardous waste from the time it is generated to its ultimate disposal. Hazardous waste management planning considers the present hazardous waste management as well as the anticipated future management. Hazardous waste management planning consists of:

1. a hazardous waste compliance program,
2. waste minimization and pollution prevention,
3. a capacity assurance plan,
4. facility siting, and
5. a state hazardous waste management plan.

The discussion of each of these plans and programs can be found in Chapter 7.

Site correction and remediation planning deals with activities aimed at cleaning up hazardous-waste-contaminated sites. The cleanup activities may take the form of removing the contaminated soils from a site or remediating the contamination on-site. This planning is discussed in Chapter 8.

Emergency response and hazard management planning is concerned with the safety of people and the environment from potential hazards resulting from hazardous waste accidents, spills, or improper management of hazardous waste. The planning activity examines potential hazards that occur in a facility. This is discussed in detail in Chapter 9.

Citizen participation planning represents a component in hazardous waste planning because it is essential for the success of planning activities. Public participation is also required by hazardous waste laws. Chapter 10 addresses this planning.

Part Two — Legal Aspects

The two chapters of Part Two describe the legal background in terms of the laws, regulations, rules, and procedures related to the management of hazardous waste. Chapter 3 describes applicable federal laws and regulations for the hazardous waste program. It focuses on Subtitle C of the Resource Conservation and Recovery Act (RCRA) and its implementing regulations. Also discussed in this chapter is the Comprehensive Environmental Response, Compensation, and Liability Act (CERCLA) and its relevance to the management of hazardous waste.

Chapter 4 discusses rule development and the legal procedures that delineate the authority between the EPA and the states, and it examines how this separation of authority affects the regulated industry. The chapter describes the rationale and the process of state authorization to implement the RCRA program and analyzes the RCRA rulemaking process at the state level.

3 Statutory and Regulatory Background

As discussed in Chapter 1, after World War II there was a growing national concern that our ability to manage solid waste was insufficient and needed improvement. Most wastes were disposed of in landfills or open dumps. The latter practice is particularly harmful to human health because it contaminates drinking water from underground and surface supplies, and it pollutes land and air. Alternatives to this unsafe practice for waste disposal were needed.

In response to pressing public concern, Congress passed the Solid Waste Disposal Act (SWDA) in 1965, which provided technical and financial assistance to states, local governments, and interstate agencies for the development of a solid waste management plan. It also provided training grants for the design, operation, and maintenance of solid waste disposal systems. Although the Act expressed concerns over hazardous waste management practices, it was passed with the primary purpose of improving solid waste disposal practices.

The SWDA defines hazardous waste as a solid waste (or combination of solid wastes) which, because of its amount, concentration, or physical, chemical, or infectious characteristics, may contribute to adverse health effects; and when improperly treated, stored, transported, disposed of, or otherwise managed, it may pose a hazard to human health and the environment. The definition indicates that hazardous waste is a specific form of solid waste. Physically, solid wastes may be either solid or liquid, or may contain gas.

Amendments to the SWDA raised solid waste management to a higher level of sophistication. The amendments to the Act are the Resource Recovery Act (RRA) of 1970 and the Resource Conservation and Recovery Act (RCRA) in 1976. The changes embodied in the RCRA greatly expanded the provisions pertaining to the management of hazardous waste.

The RCRA has been amended twice since 1976: in 1980 and in 1984. The 1984 amendments, called the Hazardous Waste and Solid Waste Act (HSWA), expand both the scope and detail of the requirements of the RCRA. The HSWA has added many new requirements which represent a challenge to the aspiration and imagination of the regulated community (Case, 1993).[85] These will include (1) the use of innovative and emerging technologies to improve the treatment of hazardous waste, and (2) the changing of raw materials or modifications to production processes to prevent or reduce the generation of hazardous waste. Although the term RCRA is the acronym for the Resource Conservation and Recovery Act, it is often used to refer to the overall program resulting from the Act and its amendment.

The following sections will provide readers with discussions on the scope and goals of RCRA, the contents of RCRA regulations, and the process of promulgating new RCRA regulations. Planning provides goals and objectives. Without legal effectuation tools, planning would accomplish little in achieving goals and objectives (Smith, 1983).[86] This implies that in the area of hazardous waste management, RCRA statutes and regulations represent the primary tools for hazardous waste planning.

Knowledge of hazardous waste regulations is vital for planners, because the regulatory aspect of RCRA directly affects planning. RCRA regulations outline the procedures necessary to perform the proper handling of hazardous wastes. The provisions specify what materials are regulated as hazardous wastes, what businesses are required to comply with the regulations, and in what way and under what operating conditions compliance should occur.

RESOURCE CONSERVATION AND RECOVERY ACT

RCRA has ten subtitles, labeled A through J (Table 3.1). Of the subtitles listed in Table 3.1, subtitles D, I, and C outline the framework for the three major programs that make up RCRA:

- Under Subtitle D, states are encouraged to develop comprehensive plans for the management of solid waste.
- The program described under Subtitle I relates to underground storage tanks (UST).
- Subtitle C of RCRA establishes a management system of hazardous waste from the time it is generated until its ultimate disposal, the cradle-to-grave approach.

Although hazardous waste is by definition part of solid waste, Subtitle D of RCRA does not regulate hazardous waste. It specifically refers only to the non-hazardous portion of solid waste. Subtitle C is the statute that controls the management of hazardous waste. Subtitle I was created by Congress under the HSWA to control and prevent release from USTs storing petroleum and other hazardous substances, but it does not regulate tanks storing hazardous waste. Such tanks are regulated under Subtitle C.

When RCRA is mentioned, most people think of Subtitle C. In other words, RCRA has always been associated with hazardous waste even though the association is not entirely accurate.

RCRA describes the kind of waste management program that Congress wants to establish; RCRA also provides the EPA, or the EPA's authorized states, with

Table 3.1 List of RCRA Subtitles

Subtitle A	General provisions
Subtitle B	Office of Solid Waste; authorities of the administrator
Subtitle C	Hazardous waste management
Subtitle D	State or regional solid waste plans
Subtitle E	Duties of the Secretary of Commerce in resource and recovery
Subtitle F	Federal responsibilities
Subtitle G	Miscellaneous provisions
Subtitle H	Research, development, demonstration, and information
Subtitle I	Regulation of underground storage tanks
Subtitle J	Demonstration program for medical waste tracking

the authority necessary to carry out the intent of the Act. The goals set by RCRA are straightforward. They include:

1. the protection of human health and the environment,
2. the reduction of waste and the conservation of energy and natural resources, and
3. the reduction or elimination of the generation of hazardous waste as expeditiously as possible.

The RCRA program is made up of several pieces, including the promulgation of RCRA regulations and the issuance of RCRA guidance and policy. RCRA regulations are the legal mechanisms that spell out how the broad policy directives of the Act are to be implemented. RCRA guidance documents are issued by the EPA primarily to elaborate and provide direction on the implementation of RCRA regulations. They are also used to provide the EPA's interpretation of the act and their regulations. RCRA policy statements specify the EPA's operating procedures that must be followed (U.S. EPA, 1990).[87]

RCRA REGULATIONS

RCRA provides a Congressional mandate for the development of RCRA regulations. These regulations are legal mechanisms that describe how RCRA's objectives should be achieved. Many RCRA regulations have already been developed, and many more are expected to be. RCRA regulations are continuously being developed and amended to accommodate societal needs for better hazardous waste management, safer facility operation, and a healthier environment.

The procedure for promulgating a new regulation starts with the publication of a proposed regulation in the **Federal Register** (FR). The register is an official publication of the federal government specifically dealing with aspects of regulation development. The FR is published every working day year-round. It notifies the public of proposed and final regulations promulgated by all federal agencies.

RCRA regulations are proposed by the EPA. The wording of such a proposed regulation includes a discussion of the EPA's rationale for proposing the regulation. This discussion is placed in the preamble of the proposed regulation.

To retrieve a proposed or a newly promulgated regulation from the FR, we have to know the date and the first page number of the regulation in the register. The commonly used notation for any reference in the FR is "(volume) FR (page)." The regulation on the listing of spent solvent, for example, can be found in the 50 FR 53,315. This means that this specific regulation is published in the FR volume 50, starting from page 53,315. Volume 50 refers to the year 1985, volume 51 the year 1986, volume 56 the year 1991, and so on. The FR is available in most public libraries.

Once the proposed regulations are published in the register, the EPA allows the public to comment on them — normally within 60 days. It is important for industry, citizen groups, and the general public to utilize this public comment period to express their concerns to the EPA. Following the comment period, the EPA will revise the proposed regulation to accommodate public comments and will finalize the regulation through publication in the FR. The EPA may also drop the proposed regulations based on comments the agency receives.

Annually on the first day of July, the newly promulgated regulations are added to the existing regulations in the **Code of Federal Regulations** (CFR) in a highly structured format. This process is called codification. Normally, CFR documents are available in most public libraries. Sales of CFR documents are handled exclusively by the Superintendent of Documents, U.S. Government Printing Office, Washington, D.C. 20402. The telephone number is (202) 783-3238.

Once published, the new regulations are allocated systematically to appropriate agencies. Hazardous waste regulations are compiled in Volume 40 of the CFR in Parts 260 through 272. These regulations are cited as 40 CFR, with the part and section number listed afterward. The citation that read 40 CFR Part 262.41 means volume 40 of the CFR, Part 262, Section 41.

The content of the RCRA regulations in 40 CFR is arranged according to the order shown in Table 3.2. The order demonstrates that the RCRA program affects three major business entities: the generators of hazardous waste; the transporters; and the treatment, storage, and disposal (TSD) facilities. These hazardous waste handlers represent the overall RCRA-regulated community.

Some of the provisions that are important to hazardous waste handlers are (1) waste determination, (2) notification, (3) hazardous waste generator requirements, (4) transporter requirements, (5) TSD facility requirements, and (6) permits. The following sections briefly describe these components.

Waste Determination

The generators are the first link in the cradle-to-grave chain of a hazardous waste management system. It is the responsibility of the generator of the waste to determine if the waste produced is indeed hazardous. In this waste determination

Table 3.2 Parts 260 through 272 of 40 CFR

Part 260 —	Hazardous Waste Management System
Part 261 —	Identification and listing of hazardous waste
Part 262 —	Standards applicable to generators of hazardous waste
Part 263 —	Standards applicable to transporters of hazardous waste
Part 264 —	Standards for owners/operators of hazardous waste treatment, storage, and disposal (TSD) facilities
Part 265 —	Interim status standards for owners/operators of TSD facilities
Part 266 —	Standards for the management of specific hazardous wastes
Part 267 —	Interim standards for owners/operators of new land disposal facilities
Part 268 —	Land disposal restrictions
Part 269 —	(reserved)
Part 270 —	The hazardous waste permit program
Part 271 —	Requirements for authorization of state programs
Part 272 —	Approved state programs

phase, the generator must first determine if the waste is a solid waste. Regulations regarding the solid waste definition are in 40 CFR 261.2.

A **solid waste** under 40 CFR 261.2a(1) is defined as any discarded material that is not qualified for either solid waste exclusions or for recycling exclusions. 40 CFR 261.4(a) specifies certain materials (such as domestic sewage) that are excluded from the definition of solid waste. Also excluded in the definition are certain recycled materials (such as materials that have been reclaimed but must be reclaimed further before they are completely recovered) that are specified in 40 CFR 260.30 and 260.31.

To clarify the solid waste definition, RCRA regulations define a **discarded material** in 40 CFR 261.2a(2) as any material that is *abandoned* (such as material being disposed of), *recycled* (such as material burned for energy recovery), or *considered inherently waste-like* (such as recycled reactant wastes of hexachlorobenzenes under alkaline conditions).

Once a material is found to be a solid waste, the next question is whether it is a hazardous waste. Regulations regarding the hazardous waste definition and exclusions are found in 40 CFR 261.3 and 261.4. Under 40 CFR 261.3, a solid waste is defined as a **hazardous waste** if it is not excluded from regulation as a hazardous waste. An example of this exemption is household hazardous waste. This waste is definitely hazardous, but legally it is classified as solid waste.

If a solid waste does not qualify for an exemption, it will be deemed a hazardous waste when the waste either:

1. is listed by the EPA in 40 CFR 261.30 through 261.33, or
2. exhibits any of the four hazardous waste characteristics (ignitability, corrosivity, reactivity, and toxicity) described in 40 CFR 261.20 through 261.24.

This determination process results in the designation of a hazardous waste as a *listed* or *characteristic* hazardous waste. A **listed hazardous waste** consists of the following groups:

1. Nonspecific source wastes (40 CFR 261.31) are commonly generated by manufacturing and industrial processes. An example from this list is spent halogenated solvents used in degreasing. The EPA has coded this hazardous waste as F001. Another example is wastewater treatment sludge from electroplating processes, which is coded as F006. EPA codes for nonspecific source wastes always start with "F"; therefore, nonspecific source wastes are popularly known as the **F-wastes**.
2. Specific source wastes (40 CFR 261.32) are generated from specifically identified industries such as wood preserving, petroleum refining, and organic chemical manufacturing. The wastes include sludges, still bottoms, wastewaters, spent catalysts, and residues. They are known as **K-wastes**. Examples are bottom-sediment sludge from the treatment of wastewaters from wood-preserving processes (K001) and oven residue from the production of chrome oxide green pigments (K008).
3. Discarded commercial chemical products (40 CFR 261.33[e] and 261.33[f]) consist of acute hazardous waste (**P-wastes**) and toxic hazardous waste (**U-wastes**). Examples are arsenic acid (P010) and xylene (U239).

Characteristic hazardous wastes are commonly called **D-wastes**. These wastes are determined either by the generator through general knowledge based on daily operation, or through laboratory testing. D001 represents *ignitable* waste (40 CFR 261.21). A solid waste that exhibits any of the following properties is considered a hazardous waste due to its **ignitability**:

- a liquid — other than aqueous solutions containing less than 24 percent alcohol — that has a flashpoint less than 60°C or 140°F;
- a nonliquid that, under standard temperature and pressure, is capable of spontaneous and sustained combustion;
- an ignitable compressed gas, specified by the Department of Transportation (DOT) under 49 CFR 173.300; and
- an oxidizer per DOT regulations of 49 CFR 173.51.

Under 40 CFR 261.22, the EPA specifies the criteria for a *corrosive* waste. D002 is used for coding such waste. A solid waste that exhibits any of the following properties is considered a hazardous waste due to its **corrosivity**:

- an aqueous material with a pH less than or equal to 2, or greater than or equal to 12.5; or
- a liquid that corrodes steel at a rate greater than 0.25 in. or 6.35 mm per year at a temperature of 55°C or 130°F.

D003 is a *reactive* waste under 40 CFR 261.23. A solid waste that exhibits any of the following properties is considered a hazardous waste due to its **reactivity**:

- is normally unstable and readily undergoes violent change without detonating;
- reacts violently with water;
- forms an explosive mixture with water;
- generates toxic gases, vapors, or fumes when mixed with water;
- contains cyanide or sulfide and generates toxic gases, vapors, or fumes at a pH between 2 and 12.5;
- is capable of detonation if heated under confinement or if subjected to a strong initiating source;
- is capable of detonation at standard temperature and pressure; or
- is listed by the DOT as explosive in 49 CFR.51, 53, and 88.

A solid waste that exhibits the characteristic of *toxicity* is coded by 40 CFR 261 as D004 through D043. The waste may have *metal* constituents (D004 through D011), *insecticides* (D012 through D015), *herbicides* (D016, D017), or *organics* (D018 through D043). The **toxicity** characteristics are tested using the Toxicity Characteristics Leaching Procedure (TCLP) test, which is designed to identify wastes likely to leach hazardous concentrations of particular toxic constituents into surrounding soils or groundwater as a result of improper management of hazardous waste. The procedure extracts constituents from the tested waste to simulate the leaching actions that occur in landfills. The extract is then analyzed to determine if it possesses any of the TCLP contaminants specified in 40 CFR 261.24(b). If the concentrations of the toxic constituents exceed the levels listed in 40 CFR 262.24(b), the waste is classified as hazardous. The list of EPA waste codes, TCLP contaminants, and the maximum concentration of contaminants (in mg/L) of each contaminant are provided in Table 3.3.

In addition to the *listed waste* and *characteristic waste,* which altogether represents more than 700 waste codes, the EPA also defines certain other wastes based on what are called the *mixture rule, derived-from rule,* and the *contained-in rule.* Under the **mixture rule**, a mixture of a listed waste and a solid waste must be considered a hazardous waste unless the mixture qualifies for an exemption (40 CFR 261.3[a][2]). Under the **derived-from rule**, a waste that is generated from the treatment, storage, or disposal of a hazardous waste (such as ash, leachate, or emission control dust) is also a hazardous waste, unless it is exempted (40 CFR 261.3[b]). Under the **contained-in rule**, soil, groundwater, surface water, or debris that is contaminated with listed waste must be regulated as hazardous waste (40 CFR 261.3[f]).

If a generator determines that the waste generated is hazardous, the company must comply with RCRA regulations. As an initial step, the generator must file a notification with the EPA using EPA Form 8700-12. Upon receipt of the form, the EPA will assign an EPA identification number to the generator.

Table 3.3 Maximum Concentration of Contaminants for the Toxicity Characteristics (40 CFR 261.24)

Waste code	Contaminant	Regulatory level (mg/L)
D004	Arsenic	5.0
D005	Barium	100.0
D018	Benzene	0.5
D006	Cadmium	1.0
D019	Carbon tetrachloride	0.5
D020	Chlordane	0.03
D021	Chlorobenzene	100.0
D022	Chloroform	6.0
D007	Chromium	5.0
D023	o-Cresol	200.0[a]
D024	m-Cresol	200.0[a]
D025	p-Cresol	200.0[a]
D026	Cresol	200.0[a]
D016	2,4-D	10.0
D027	1,4-Dichlorobenzene	7.5
D028	1,2-Dichloroethane	0.5
D029	1,1-Dichloroethylene	0.7
D030	2,4-Dinitrotoluene	0.13[b]
D012	Endrin	0.02
D031	Heptachlor (and its epoxide)	0.008
D032	Hexachlorobenzene	0.13[b]
D033	Hexachlorobutadiene	0.5
D034	Hexachloroethane	3.0
D008	Lead	5.0
D013	Lindane	0.4
D009	Mercury	0.2
D014	Methoxychlor	10.0
D035	Methyl ethyl ketone	200.0
D036	Nitrobenzene	2.0
D037	Pentachlorophenol	100.0
D038	Pyridine	5.0[b]
D010	Selenium	1.0
D011	Silver	5.0
D039	Tetrachloroethylene	0.7
D015	Toxaphene	0.5
D040	Trichloroethylene	0.5
D041	2,4,5-Trichlorophenol	400.0
D042	2,4,6-Trichlorophenol	2.0

Table 3.3 Maximum Concentration of Contaminants for the Toxicity Characteristics (40 CFR 261.24) (Continued)

Waste code	Contaminant	Regulatory level (mg/L)
D017	2,4,5-TP (Silvex)	1.0
D043	Vinyl chloride	0.2

[a] If o-, m-, and p-cresol concentrations cannot be differentiated, the total cresol (D026) concentration is used. The regulatory level of total cresol is 200 mg/L.

[b] Quantification limit is greater than the calculated regulatory level. The quantification limit therefore becomes the regulatory level.

Notification Requirement

The notification requirement is applicable to hazardous waste generators (40 CFR 262.12) as well as to the transporters (40 CFR 263.11) and TSD facilities (40 CFR 264.11 and 265.11). Having an EPA ID number is, in fact, mandatory for all hazardous waste handlers. Without this EPA ID number, the handler is barred from storing, transporting, or offering for transportation, treating, and disposing of any hazardous waste. Also, a generator is forbidden from offering his or her hazardous waste to any transporter or TSD facility that does not have an EPA ID number.

A business that is subject to the notification requirement should submit one notification form per site or location to the EPA. If a business conducts hazardous waste activities at more than one location, it must submit a separate form for each location.

The two-page notification form consists of several items. It includes the name of the business, mailing address, facility location, name and phone number of contact person, ownership, and information on hazardous waste types and waste handling activities — i.e., whether the handler is a generator, transporter, or TSD facility. After the form is completed, dated, and signed, it is ready for submittal to the EPA.

The notification requirement represents a starting point for the EPA to systematically track the movement of every hazardous waste from the generator to its ultimate disposal facility. Notification is not, however, the only requirement for hazardous waste generators. They still are subject to other RCRA regulations as described below.

Hazardous Waste Generator

In general, the generator has the responsibility of ensuring the proper storage and handling of his hazardous waste. 40 CFR 260.10 defines a generator as any person, by site, whose act or processes produces hazardous waste identified or listed in 40 CFR 261, or whose act first causes a hazardous waste to become subject to regulation.

A generator may accumulate hazardous waste on-site without a hazardous waste storage permit for 90 days or less (40 CFR 262.34). The 90-day period provision allows a generator to collect enough waste for one big shipment in order to make transportation more cost-effective. In addition to proper storage practices, the generator must ensure that facility personnel have adequate training in the proper handling of hazardous waste, and that the generator has developed a written contingency plan addressing emergency response situations (40 CFR 262.34[a][4]).

To transport hazardous waste off-site, the generator must ensure proper packaging and proper labeling, marking, and placarding of the packaged waste (40 CFR 262.30 through 262.33). The purpose of proper packaging is to prevent leakage of hazardous waste during the normal course of transportation and during unexpected situations such as a traffic accident or a waste-handling mishap (e.g., a hazardous waste drum dropped from a truck bed). Labeling, marking, and placarding of the packaged waste is done to inform the shipping crew, firefighters, and emergency responders about the characteristics and dangers associated with the waste being transported. This information is vital in case of an emergency situation.

The generator must prepare a shipping document called the Uniform Hazardous Waste Manifest (EPA Form 8700-22), which accompanies the hazardous waste at all times. Normally, the manifest can be obtained from the state environmental agency (40 CFR 262.20 and 262.21).

The generator specifies on the manifest the name and EPA ID numbers of each transporter, the designated TSD facility that will receive the waste, the type of waste, the amount, certification that it is properly packaged and labeled, and his handwritten signature on the multi-copy manifest certifications. Sufficient copies of the manifest must be prepared so that all handlers listed on it will be provided a copy (40 CFR 262.22 and 262.23).

After the transporter signs the manifest, the generator retains a copy of the manifest and gives the remaining copies of the manifest to the transporter. Upon arrival of the shipment at the TSD facility, the transporter retains a copy and gives the remaining copies to the TSD facility. The TSD retains one copy and sends a copy of the manifest — which has the signatures of the generator, transporter, and TSD facility — to the generator. The generator must keep the copy for at least 3 years (40 CFR 262.40).

If the generator does not receive the copy of the manifest with all of the signatures within 35 days from the date the waste was accepted by the transporter, the generator is required to trace the whereabouts of the shipment. If the generator does not receive the copy of the manifest with all of the signatures within 45 days, the generator must file an exception report to the EPA detailing the efforts the generator has made to locate the waste.

Generators are subject to extensive record-keeping and reporting requirements. By March 1 of even-numbered years for the preceding calendar year, a generator must file biennial reports with the EPA or an authorized state using EPA Form 8700-13A (40 CFR 262.41).[88] The report identifies each transporter used and each TSD facility to which wastes were shipped, and it describes the type and amount of each type of waste handled by the generator. The generator must maintain copies of the reports for at least 3 years (40 CFR 262.40).

The above requirements apply to generators producing at least 2,200 lb of hazardous waste, or 2.2 lb of acutely hazardous waste, in any month of a calendar year. This type of generator is known as a **large-quantity generator** (LQG). A **small-quantity generator** (SQG) generates between 220 and 2,200 lb of hazardous waste, or less than 2.2 lb of acutely hazardous waste, in any calendar month. An SQG is not required to comply with the biennial reporting requirement and may store the waste on-site for up to 180 days (40 CFR 262.34[d] and 40 CFR 262.44). A generator that produces less than 220 lb of hazardous waste in any month is called a **conditionally exempt small-quantity generator** (CESQG) and is only subject to certain minimum standards (40 CFR 261.5).

Transporter of Hazardous Waste

Transporter requirements apply to anyone who moves a hazardous waste off the site from where it was generated. This means that the provision is not applicable to generators who engage in on-site transportation of their hazardous waste (40 CFR 263.10). A hazardous waste produced by a generator (not by a TSD facility) is categorized as a **primary hazardous waste**. A TSD facility may become a generator if the facility also produces a **(secondary) hazardous waste** as a result of a hazardous waste treatment process within the facility. This TSD is subject to the generator requirements in addition to applicable TSD standards. It is not uncommon to find a TSD facility that is a generator *and* a transporter of hazardous waste.

A transporter can only accept hazardous waste that is accompanied by a manifest signed by the generator (40 CFR 263.20). The transporter who accepts manifested hazardous waste must sign and date the manifest acknowledging acceptance of the waste, and he must return one copy to the generator before leaving the generator's property. The transporter is responsible for the shipment until the cargo arrives at the TSD facility. When the transporter delivers the waste to the designated TSD facility, the transporter must date the manifest and obtain the signature of the TSD operator, retain one copy of the manifest for his own records, and give the remaining copies to the person receiving the waste. The transporter must keep the copy of the manifest for at least 3 years (40 CFR 263.22). If the waste cannot be delivered in accordance with the manifest, the transporter must contact the generator for further instructions and revise the manifest according to the generator's instructions (40 CFR 263.21).

The transporter may hold a hazardous waste in containers for up to 10 days at a hazardous waste transfer facility without being required to obtain an RCRA storage permit (40 CFR 263.12). Transfer facilities include loading docks, parking areas, and storage areas where shipments are held during the normal course of transportation (40 CFR 260.10).

If an accidental spill or leakage occurs during transportation, the transporter is responsible for its cleanup (40 CFR 263.30). The transporter must take immediate response action to protect human health and the environment, such as containment of the spill and notification of local police and fire departments.

Because of the potential liabilities from accidental spills, leaks, improper handling, and illegal disposal during transportation, generators prefer to maintain control of their wastes by transporting them in company-owned trucks (Wentz, 1989).[89] Alternatively, Wentz observes, generators feel more confident that their wastes will reach the designated TSD if they are picked up and transported by the TSD's trucking operation.

TSD Facility

As far as a TSD facility is concerned, the RCRA program sets both (1) the technical standards for the design and safe operation of TSD facilities, and (2) the requirements for issuing the permits that each TSD facility is required to have. The technical standards and the permitting process are interrelated in that the permitting process enforces the application of technical standards to TSD facilities.

40 CFR 260.10 describes a TSD facility as follows:

- A facility is regulated as a *treatment* facility if it utilizes any method or process designed to change the physical, chemical, or biological character or composition of hazardous waste so as to neutralize such waste, to recover material or energy from the waste, or to render the waste nonhazardous or less hazardous, safer to handle, or reduced in volume.
- A *storage* facility is a facility that engages in the holding of hazardous waste for a temporary period; at the end of that period, the waste is treated, disposed of, or stored elsewhere.
- A *disposal* facility is a facility where hazardous waste is intentionally placed into or on land or water, where waste will remain after closure of the facility.

Separate requirements have been issued for **permitted TSD** facilities (40 CFR 264) and **interim status TSD** (40 CFR 265). Both Parts 264 and 265 include specific provisions for each hazardous waste management type and general requirements. Examples of hazardous waste management types are landfill, surface impoundment, incinerator, tank, and container.

- An interim status facility is allowed to continue its operation without permit. The facility must be in existence prior to November 19, 1980, must notify the EPA of its operations, and must file a permit application.

- Permitted facilities include new TSD facilities and existing facilities that did not qualify for interim status, both of which must first obtain RCRA permits to begin or continue operations.

General requirements are similar for both permitted and interim status TSDs. They include the following provisions:

1. A TSD must conduct a detailed chemical and physical analysis of a representative sample of a hazardous waste before the waste is treated, stored, or disposed of at the facility (40 CFR 264.13 and 265.13).
2. A TSD must install a security system (such as a fence or a surveillance system) to prevent unknowing entry of persons or livestock onto the active portion of the facility (40 CFR 264.14 and 265.14).
3. A TSD is required to prepare and implement an inspection plan for malfunctions and deterioration, operator errors, and discharges that may be causing release of hazardous waste constituents into the environment or threatening human health (40 CFR 264.15 and 265.15).
4. TSD personnel are required to be properly and annually trained in hazardous waste management procedures relevant to the positions in which they are employed. Personnel training should include emergency preparedness and response procedures (40 CFR 264.16 and 265.16).
5. A TSD must ensure that all ignitable or reactive wastes must be protected from sources of ignition or reaction, and that the treatment, storage, and disposal of ignitable, reactive, or incompatible waste does not result in damage to the containment structure or threaten human health or the environment (40 CFR 264.17 and 265.17).
6. A TSD is required to have specific emergency preparedness equipment (such as an alarm system, fire and spill control, and decontamination equipment) and other precautions as part of incident preparedness and spill prevention procedures (40 CFR 264.30 through 264.37 and 265.30 through 265.37).
7. A TSD is required to have an adequate contingency plan, which describes an action plan to address all types of emergency situations that might occur in the facility (40 CFR 264.50 through 264.56, and 265.50 through 265.56).
8. A TSD facility must be covered by liability insurance for claims arising out of injuries to persons that result from facility operations (40 CFR 264.143[f] and 265.143[e]).
9. Similar to the generator and transporter provisions, TSD facilities are also subject to manifest (40 CFR 264.71, 72 and 265.71, 72) and biennial reporting requirements (40 CFR 264.75 and 265.75).
10. A TSD facility is subject to closure requirements (40 CFR 264.110 through 264.116 and 265.110 through 265.116). These requirements are intended to cover a period when the TSD facility must cease

treatment, storage, or disposal operations. The facility must address this situation by having a detailed written closure plan, an implementation schedule, and a cost estimate for closure.

11. A land disposal facility (landfill, surface impoundment, waste piles, or land treatment facility) is subject to post-closure requirements (40 CFR 264.117 through 264.120 and 265.117 through 265.120). Post-closure is the 30-year period after closure when facilities must perform certain monitoring and maintenance activities including groundwater monitoring, maintenance of monitoring systems, and maintenance of waste-containment systems.
12. A land disposal facility is required to perform groundwater monitoring in order to assess the impact of a facility on the groundwater beneath it (40 CFR 264.90 through 264.101 and 265.90 through 265.94). Monitoring must continue for 30 years after the facility has closed.
13. Financial responsibility requirements have been established to ensure that any TSD facility has adequate and available funds for closure and post-closure care (40 CFR 264.140 through 264.151 and 265.140 through 265.150).

The requirements for a permitted TSD and an interim status TSD facility differ in a number of ways. A set of stricter standards to prevent the release of hazardous waste are applicable to a permitted TSD (40 CFR 264.175). When groundwater contamination is detected, or when releases of hazardous waste are found from solid waste management units (defined as a unit where any solid waste was placed), corrective actions must be taken. In this case, corrective action requirements for a permitted facility (40 CFR 264.100) are broader than those for an interim status facility (40 CFR 265.93). However, the EPA still maintains legal authority to order interim status facilities to undertake broader corrective actions outside the permitting status (Superfund Report, 1990).[90] As specified in Section 3004(u) of RCRA, the corrective action provision applies similarly to permitted facilities as well as to those facilities operating under interim status (Davis, 1993).[91]

In addition to the above 13 general requirements, the EPA has also promulgated specific design, construction, and operating standards for each part of a TSD facility. The parts regulated in the 40 CFR are:

- containers, including drums, pails, and boxes;
- tanks, any ancillary piping, and containment system;
- surface impoundments, excavation, or diked area for holding liquid hazardous wastes;
- waste piles, i.e., non-containerized accumulation of solid hazardous wastes;
- land treatment, where hazardous waste is applied onto the soil surface (such facilities are disposal facilities if the waste remain after closure);
- landfills, where hazardous waste is disposed in or on the land;

- incinerators, which destroy hazardous wastes;
- underground injection well, i.e., subsurface emplacement of liquid hazardous waste through a well;
- drip pads used to convey drippage from treated wood at wood-preserving plants; and
- treatment of hazardous waste associated with distillation, fractionation, air stripping, and other operations.

Permit

RCRA requires every TSD facility to obtain a permit. The permit application is divided into two parts, A and B.

- Part A is a shorter form that collects general information about a facility, such as name, location, nature of business, regulated activities, and a topographic map of the facility (40 CFR 270.13).
- Part B is much more extensive than Part A in that it requires the facility to supply detailed and highly technical information based on standards specified in 40 CFR Parts 264 and 270.

New facilities must submit Part A and Part B simultaneously. Interim status facilities submitted Part A when they applied for their interim status and must submit Part B applications in accordance with deadlines established by the EPA or the state. Failure to meet a deadline may result in the facility losing its interim status, and therefore the facility must close.

After a complete application is filed, it is reviewed and is either approved or denied by the EPA or the state. Procedures for processing the application and issuing the permit are described in 40 CFR Part 124. These procedures include preparation of draft permits, public comment and hearing, and the issuance of final decisions.

In many circumstances, a TSD facility may also have to comply with environmental acts other than RCRA, such as the Clean Air Act, the Clean Water Act, the Safe Drinking Water Act, or the Toxic Substance Control Act. Compliance with those laws includes obtaining the required permits.

COMPREHENSIVE ENVIRONMENTAL RESPONSE, COMPENSATION, AND LIABILITY ACT

Although RCRA creates a framework for the proper management of wastes, the Act does not address the problems of hazardous waste found at inactive or abandoned sites or those spills that require emergency response. These problems are taken care of by the Comprehensive Environmental Response, Compensation, and Liability Act (CERCLA).

CERCLA was created in response to the discovery of a large number of inactive and abandoned hazardous waste dumps in the late 1970s. Such dumps can introduce hazardous waste to the environment and endanger human health and the environment.

As discussed in Chapter 1, one of the best-known of these dumps is Love Canal near Buffalo, NY, where a chemical company had buried a large amount of hazardous waste in a dry canal. After the canal was capped with clay and soil during the 1940s and early 1950s, a community was established around it. Over the years the canal leaked, and by 1978 the threat to human health had become so great that President Carter was forced to declare the area a national disaster and relocate families in its vicinity.

When Love Canal gained national attention, the EPA had few tools to deal with a disaster of this type and magnitude. Although RCRA allowed the EPA to require proper management of hazardous waste at active and properly closed facilities and to compel persons to abate endangerments to human health and the environment, RCRA provided limited authority for governmental response to leaks or threatened leaks of hazardous waste at abandoned or inactive sites. It quickly became apparent to the President, Congress, and the public that some type of national legislation was needed to fill this void. This legislation took the form of CERCLA, which was passed by Congress in 1980 and was signed into law by President Carter.

The CERCLA program contains three objectives:

1. To take actions against releases or threatened releases of hazardous chemicals that may present a danger to the public and the environment.
2. To identify individuals or firms responsible for a contaminated site before taking any response action. These individuals and firms are defined as responsible parties (RPs). Once the RPs are identified, the EPA requires the RPs to take appropriate response action to clean up the site. The action may include the removal of contaminated soil and/or site remediation.
3. To clean up a contaminated site where the RPs are not found or where the RPs are unable to take the responsibility. The EPA can perform the necessary action using money from the Hazardous Substance Response Trust Fund, financed by a levy on crude oil, certain petroleum products, 42 chemical feedstocks, and appropriations from general revenues. Because of the magnitude of the appropriations, CERCLA is popularly known as "Superfund."

RCRA/CERCLA RELATION

RCRA and CERCLA are separate but interrelated programs. For example, when wastes from a CERCLA site are taken off-site for treatment, storage, or disposal, the receiving facility must be a RCRA TSD facility, and the waste must be handled in accordance with applicable RCRA requirements including the generator and transporter regulations.

The difference between the two programs is that RCRA handles "hazardous waste," which is narrowly and precisely defined in 40 CFR 261, while CERCLA deals with remediation caused by releases of a "hazardous substance." The definition of a hazardous substance is extremely broad and covers most chemicals. CERCLA defines hazardous substances as those listed or designated under various environmental laws. They include "hazardous wastes" under RCRA, "hazardous substances" and "toxic pollutants" of the Clean Water Act, "hazardous air pollutants" under the Clean Air Act, and "hazardous chemical substances" of the Toxic Substances Control Act.

When first introduced, the focus of CERCLA was to clean up contamination at inactive or abandoned sites. Prior to HSWA, no other environmental laws were in place to address the cleanup of active sites. The EPA's only authority for requiring RCRA facilities to conduct site cleanups in case of contamination was under CERCLA. With the signing of HSWA, the EPA was given corrective action authority, which requires RCRA facilities to perform cleanup activities ranging from quick-fix measures to full cleanups of their contaminated sites.

Under CERCLA, the EPA has discretion in deciding the level of cleanup as long as the cleanups permanently and significantly reduce the volume, toxicity, or mobility of the hazardous substances. However, permanent cleanups are extraordinarily expensive and difficult to accomplish (Hope, 1993).[92] Regarding active RCRA sites, the EPA's philosophy is that these sites are different from CERCLA sites in that they represent a controlled situation. This "controlled situation" designation implies that a facility can perform prompt cleanup to levels consistent with current use, but can defer final cleanup as long as the facility remains under an RCRA permit (Hall et al., 1991).[93]

Parallel to the CERCLA goal of getting the RPs to take appropriate response action to clean up a contaminated site, the act places "strict, joint, and several" liabilities on RPs associated with the site.

- **Strict liability** means that no showing of actual fault is required. An RP may claim that he was not negligent, or that his activities were consistent with standard industry practices and, therefore, he is not responsible for the site nor liable for the contamination costs. The strict liability concept does not accept this claim.
- **Joint and several liability** means that whenever there is any evidence of the co-mingling of hazardous substances by different parties, each is individually responsible for all cleanup costs at the site. This implies that multiple contributors of hazardous waste contamination at a site may all be held equally responsible unless one or more can demonstrate that its wastes can be separately identified or could not possibly have contributed to the harm.

CERCLA's strict liability has been uniformly endorsed by the courts. While CERCLA contains no statutory mandate that liability be joint and several, courts have often found such liability (Lee, 1993).[94]

The EPA has used its enforcement authority to pursue the recovery of cleanup costs and to seek judicial orders requiring RPs to perform cleanups. Potential liabilities from CERCLA provisions are clearly a major motivation for hazardous waste generators to comply with all RCRA regulations (Reinhardt, 1989).[95] Although full compliance with RCRA regulations is no defense against liability under CERCLA (Quarles, 1982),[96] the attempt is still the best defense to civil and criminal sanctions resulting from RCRA noncompliance activities (Green et al., 1993).[97]

Besides the fact that RCRA focuses its attention on current and future activities and CERCLA is principally retrospective, the major differences between RCRA and CERCLA are the role of regulations in implementing each statute and the time periods on which each statute focuses its primary attention (Briggum et al., 1985).[98] First, RCRA is not self-implementing. Congress directed the EPA to promulgate regulations that would control company activities. The primary exceptions are certain obligations created by the 1984 HSWA which do not require implementing regulations. CERCLA, on the other hand, creates a program that establishes liabilities and obligations, and does not require promulgation of regulations for implementation.

RCRA REGULATORY ASSISTANCE

RCRA is a complex program. The EPA's interpretation of the act does not end with the codification of hazardous waste regulations in 40 CFR. The EPA further clarifies its regulations through guidance documents and policy statements. Guidance documents are provided to direct the implementation of hazardous waste regulations. A typical guidance document describes, for example, how to evaluate a permit application to ensure that every requirement is included.

Policy statements are issued to outline the manner in which part of the RCRA program is to be carried out. An example of a policy statement is a description of the types of enforcement actions to be taken if a groundwater violation is found. In order to aid in implementing the RCRA program, the EPA has developed many guidance and policy documents. The agency has also set up a telephone call-in service to answer any questions regarding RCRA and CERCLA. The staff of this hotline has an up-to-date understanding of the RCRA regulatory requirements and can provide details on changes in RCRA procedural requirements. The phone number is (202) 382-3000 or toll-free (800) 424-9346.

4 Rule Development and State Authorization

Congress established the Resource Conservation and Recovery Act (RCRA) as a program that could be delegated to states. RCRA encourages states to take responsibility for program implementation and enforcement from the EPA, because the states are more familiar with the local regulated facilities and are in a better position to administer the program effectively and, at the same time, respond to local needs appropriately.

It was the intent of Congress to have the RCRA program entirely administered by states with only minimal oversight from the EPA. In order to obtain responsibility for the RCRA program, a state must develop a hazardous waste program and have it approved by the EPA. The approval procedure that states must go through is called the "state authorization" process.

From the state's perspective, obtaining **RCRA primacy**, or authorization to implement the RCRA program, is a necessity. The rationale behind this outlook is not only that the state is in a better position to respond to local needs, but also that the primacy will give the state the opportunity, authority, and credibility to protect the state's own environment through proper management of its hazardous waste program. Because of this, knowledge and understanding of the RCRA authorization process, authorization requirements, and technicalities of the process is essential for hazardous waste planners.

STATE AUTHORIZATION

In order to obtain program authorization or approval from the EPA, a state program must be equivalent to and consistent with the federal RCRA program. Consistency with the federal program is reflected in part by the state's hazardous waste laws and rules.

Under the 1976 RCRA, the states were given two options for assuming the responsibility to administer the RCRA program. They could apply for either interim status or final authorization. The requirements for authorization are specified in Title 40 of the Code of Federal Regulations Part 271 (40 CFR 271).

Any state that administered its own hazardous waste program prior to October 24, 1982, is entitled to and may request interim authorization from the

EPA to carry out that program under Subtitle C of RCRA. The interim authorization consists of Phase I and Phase II (40 CFR 271.121[b]). Phase I authorization entitles the approved state to implement the RCRA program and rules as they relate to the identification and listing of hazardous waste, generators, transporters, and interim status treatment, storage, and disposal (TSD) facilities. Phase II authorization covers the permit program for TSD facilities.

Interim authorization for Phase I and II expired on January 1, 1986 (40 CFR 271.122[b][1]). The EPA may extend the application period for good cause (40 CFR 271.122[c][1]). At the end of this period, all interim authorizations automatically expire, and the EPA shall administer the federal program in any state that has not received final authorization (40 CFR 271.122[b][2]).

The right to seek final authorization exists whether or not the state sought interim status. The final authorization application must conform to the procedural and substantive requirements established by the EPA in its implementing regulations. Provisions in 40 CFR 271.5 require that any state that seeks final authorization from the EPA must submit an application package to the EPA consisting of the following elements:

1. A letter from the governor requesting authorization.
2. Copies of applicable state rules and statutes.
3. Documentation of public participation activities.
4. An Attorney General's Statement asserting that the state has the legal mechanism in place to implement and enforce the RCRA program. The format is described in 40 CFR 271.7.
5. A program description detailing the contents of the hazardous waste program that the state wants to administer in place of the federal program as specified in 40 CFR 271.6.
6. A State-EPA Memorandum of Agreement outlining the nature of responsibilities and the level of coordination between the state and the EPA in operating the RCRA program. The outline is found in 40 CFR 271.8.

The conditions required in the application package are intended to ensure uniformity among the states. Through the application package, the state must demonstrate that its program is equivalent to the federal program, that it is consistent with the federal program, and that it provides adequate enforcement of the requirements of Subtitle C of RCRA.

In providing states with the opportunity to implement their own programs equivalent to the federal program, Congress wanted to ensure that there would be as much uniformity as possible among the states regarding how hazardous wastes are managed (U.S. EPA, 1985).[99] When final authorization of a state program is fully effective, the standards for hazardous waste handlers should be "equivalent," whether the regulatory program for hazardous waste is administered by the EPA or the state. Congress thereby intended to prevent states with less stringent programs from becoming "dumping grounds" for hazardous waste (Hall, 1991).[100]

Prior to submitting an authorization application to the EPA, the state shall issue public notice of its intent to seek program approval from the EPA and provide a time frame for comments from the public and the regulated community (40 CFR 271.20[a]).

The EPA shall make a tentative decision as to whether or not the Agency expects to grant authorization within 90 days of receiving the application package. The EPA shall give notice of its tentative decision in the Federal Register and expect comments from the public in 90 days. After the comment period is closed, the EPA shall make a final decision in the Federal Register, taking into account any comments submitted (40 CFR 271.20[d] and [e]).

To qualify for approval, a state program must also have procedures to make information regarding TSD facilities publicly available. The procedure concerning how and when to release information and the type and amount of information released must conform to Freedom of Information Act procedures.

Approved state programs are dynamic. A program status is subject to changes due to the adoption of new regulatory requirements and in response to changes in the federal program. In light of the provision that state-approved programs must be equivalent to the federal program that the state wants to administer, states must periodically (usually annually) revise their programs to incorporate these changes by a specified date.

Ordinarily, states have at least one year to make appropriate regulatory changes and two years if statutory changes are necessary. The routine authorization updates — and the legal consequences of those updates — will certainly impact the state.

IMPACTS ON A STATE

When Congress amended RCRA in 1984 and enacted the amendments as HSWA, they believed that the HSWA provisions were important to implement quickly. Therefore, EPA regulations that implement the HSWA provisions take effect in authorized states on the same day that they take effect under the federal program. Prior to HSWA, changes to federal requirements were not enforced in authorized states until the state's program was modified to incorporate such changes.

In terms of authorization status, three types of states are recognized. They are:

1. a non-authorized state,
2. an authorized state for non-HSWA provisions, and
3. an authorized state for HSWA provisions.

In practice, a state may be authorized for a combination of non-HSWA and some HSWA provisions. It is not uncommon to find that a state may only have non-HSWA authorization while its neighboring state may have already obtained non-HSWA authorization and a portion of HSWA authorization.

In states *without HSWA authorization,* the EPA is responsible for implementing HSWA provisions and issuing permit provisions required by HSWA until the state takes over authority. This has meant a significant federal presence in authorized states. One of the most serious impacts of HSWA provisions is this dual permitting system. Under the system, an authorized state (only for administering non-HSWA provisions) will still conduct permit proceedings and issue permits; however, the EPA will implement the permit-related requirements that were added by HSWA (such as the corrective action requirements). The dual permitting system ceases when the state gets HSWA authorization.

A state is required to apply for authorization for the respective HSWA regulation after promulgating an equivalent state rule. Until such time, hazardous waste handlers in a state are subject to state rules pertaining to non-HSWA and the portion of the federal regulations pertaining to that HSWA provision.

For states with *non-HSWA authorization,* hazardous waste handlers in the state are only subject to state rules pertaining to non-HSWA requirements already adopted by the state. The non-HSWA regulations not yet adopted by a non-HSWA authorized state are not in effect until the state adopts those regulations. This means that the federal right of enforcement does not apply to non-HSWA requirements in non-HSWA authorized states.

Because adopting and implementing hazardous waste rules can take several months or even years, the status of each state's hazardous waste program varies yearly (Gruber, 1991).[101] As of January 1, 1994, almost all states and territories had received non-HSWA authorization except Alaska, Hawaii, Puerto Rico, Samoa, the Northern Marianas, and the Virgin Islands. HSWA authorization has been granted to approximately half of the states and territories.

It is important for a hazardous waste management planner to be clear on the authorization aspect of his state, because all hazardous waste handlers located in the state will be affected by the authorization status. It is also important to remember that although the authorization process is designed to ensure uniformity among the states, the federal program only establishes minimum standards because RCRA only mandates that state programs cannot be less stringent than the federal program. The consequence of this provision is that significant variation exists among the states as to the level of stringency of their hazardous waste programs.

Variation in the authorization status is also caused by the different scheduling in processing the authorization package among states. Regardless of the scheduling differences, the authorization process itself is basically uniform.

AUTHORIZATION PROCESS

The EPA has published guidance manuals which provide direction to states on developing authorization applications. States absent from initial authorization should refer to the 1988 *State Consolidated RCRA Authorization Manual* (SCRAM)[102] for detailed guidance on application for the basic RCRA program. For states with initial authorization, the EPA has developed the 1990 *State Authorization Manual* (SAM)[103] to address revisions to each state's existing RCRA program.

We know that states must continue to maintain equivalency to the federal programs; therefore, modifications to the federal program may necessitate subsequent modifications to authorized state programs. The 1990 SAM, essentially a revised format of the 1988 SCRAM, can be put in a loose-leaf binder so that updated and new materials can be added as the RCRA program continues to change.

The EPA Headquarters and the regional offices review authorization applications with each region serving as the primary contact with certain states. Communication is crucial in this process. All three entities must work as a team to resolve authorization issues as they arise so that authorization may be granted to qualified states in a timely manner.

Early consultation with the EPA is necessary if a state plans to modify its program. This pre-application consultation is vital, especially if those changes modify the state's legal authorities. The purpose of this consultation is to provide the state with an early indication of potential authorization problems so they may be corrected more easily. If necessary, the EPA can assist the state in reviewing the state's draft rules, bills, and policies.

After consultation, the next step is for the state to complete a draft authorization application. Although a draft is not formally required, states are encouraged to submit one to facilitate resolution of problems prior to adoption of new state rules. The EPA headquarters and the regional office will review the draft concurrently to ensure that EPA concerns are identified at the draft stage. Comments from the EPA headquarters and the regional office are consolidated by the region, systematically numbered, and sent to the state.

Subsequent to receiving the EPA's comments on its draft, the state should start addressing those comments in its official application. It is helpful if the state explicitly responds to each EPA comment. Independently, the EPA headquarters and the regional office will review the official application. The EPA regional office will make a tentative determination of the application after any outstanding issues have been resolved. The determination to approve or disapprove the program revision is published in the Federal Register.

Part 271 of 40 CFR identifies the minimum requirements for states to obtain authorization to administer the RCRA program, therefore 40 CFR 271 will be of interest to hazardous waste planners who must deal with the authorization process. This part of the regulation does not contain any provisions that are directly applicable to generators, transporters, or TSD facilities. However, 40 CFR 271.1 will be helpful for hazardous waste planners in the private sector, because Tables 1 and 2 in 40 CFR 271.1 help determine whether a state or the EPA will administer a specific provision of the RCRA regulations (McCoy and Associates, 1992).[104]

The two important components of the authorization application from a planning perspective are the program description and the State-EPA Memorandum of Agreement (MOA). The reason they are important is because they reflect the scope of a state's program and the coordination requirement for managing the program.

Program Description

As we know, any state seeking to administer Subtitle C of the RCRA program must submit a description of the program it proposes to administer in lieu of the federal program. 40 CFR 271.6 outlines the requirements of an RCRA program description. The requirements are:

1. a narrative description of the scope, structure, coverage, and processes of the state program;
2. a description of the organization and structure of the state agency having responsibility for administering the program;
3. a description of applicable state procedures, including permitting procedures and any state administrative or judicial review procedures;
4. copies of the permit forms, application forms, and reporting forms that the state intends to employ in its program. Forms used by the state need not be identical to the forms used by the EPA but should require the same basic information. One exception to this provision is the manifest form; use of EPA Manifest Forms 870022 and 870022A is mandatory for the state RCRA program;
5. a complete description of the state's compliance tracking and enforcement program;
6. a description of the state manifest tracking system and the procedures the state will use to coordinate information with other approved state programs and the federal program regarding interstate and international shipments;
7. an estimate of the number of generators, transporters, and TSD facilities, a brief description of the types of TSD facilities, and an indication of the permit status of these facilities; and
8. an estimate of the annual quantities of hazardous waste that is generated within the state, transported into and out of the state, and stored, treated, or disposed of within the state.

In the authorization manual, the EPA provides a checklist that should be used to review a program description (PD). The checklist is very useful for hazardous waste planners, because it is designed to ensure that a PD adequately describes the organization and management of the state program, discusses the differences between the state and federal programs, describes how the state will administer and enforce its program, and demonstrates that the state program meets the tests for authorization.

The manual also provides specific directions on how to write the State-EPA MOA. The directions include what basic elements must be described in the agreement. This will be discussed in the following paragraphs.

State-EPA Memorandum of Agreement

Submission of a State-EPA MOA as part of the RCRA authorization process is required under 40 CFR 271.8. The MOA shall be executed by the state environmental agency Director and the EPA Regional Administrator. The agreement shall become effective when approved by the Regional Administrator. The

Administrator will not approve any MOA that contains provisions which restrict the EPA's statutory oversight responsibility.

The following areas of agreements between the EPA and the state must be included in the MOA:

1. the procedures for the assignment of EPA identification numbers for new generators, transporters, and TSD facilities;
2. the frequency and content of reports, documents, and other information that the state is required to submit to the EPA. The state shall allow the EPA to routinely review state files relevant to the administration and enforcement of the approved program;
3. the coordination of the compliance-monitoring activities by the state and the EPA. These may specify the basis upon which the EPA will select facilities within the state for EPA inspection;
4. the EPA's compliance inspections of all hazardous waste handlers in each year for which the state is operating under final authorization;
5. the EPA's right to inspect any hazardous waste handler which it believes is not in compliance with RCRA. Usually, before conducting such an inspection, the EPA will allow the state a reasonable opportunity to conduct a compliance evaluation inspection;
6. the transfer from the EPA to the state of pending permit applications and any other information relevant to program operation not already in the possession of the state;
7. the specification of classes and categories of permit applications, draft permits, and proposed permits that the state will send to the EPA for review and comment;
8. the joint processing of permits by the state and the EPA for facilities that require permits from both the EPA and the state;
9. the submission from the state to the EPA of copies of draft permits and permit applications for all major facilities for review and comment;
10. the state's review of all permits issued under state law prior to the date of program approval. These permits may be modified, or revoked and reissued; and
11. the revision of the MOA in order to maintain consistency with the annual program grant and the State-EPA agreement. In an effort to support state programs, the EPA awards an RCRA grant to the states every year. Detailed program priorities and specific arrangements for EPA support of a state program will change and are more appropriately negotiated in the context of annual agreements rather than in the MOA.

To expedite the authorization process, the EPA has provided in the 1990 SAM a model of the MOA for use by hazardous waste planners in the authorization application package. The structure of the model shows that MOAs are state-specific and that no two MOAs are exactly alike.

Once the MOA and other authorization documents are finalized, the authorization package is ready to be transmitted to the EPA. The EPA's approval of the application depends on whether the state program is equivalent to the federal program. This is reflected by, among other things, the scope and content of the state rules.

RULE DEVELOPMENT

Various states have adopted RCRA regulations in order to maintain the equivalency requirement. The effort can be done either by (1) reprinting federal RCRA regulations in their Administrative Codes using appropriate language and format, or (2) by adopting the federal regulations by reference.

Depending on the state's rulemaking procedures, the rule development activity normally consists of informal and formal processes. As an example, Arizona's current rulemaking process is described below. In Arizona, the informal process covers the following steps:

1. Conduct research. This step includes the state environmental agency's review of the newly promulgated federal RCRA regulations, which can be found in the Federal Register. Additionally, a number of computer software programs are currently available that can be used for tracking new regulations.
2. Develop concept paper. The concept paper summarizes the contents of the new federal regulations and explains how they differ from the existing state rules. Once written, the paper would be circulated to RCRA compliance enforcement staff such as hazardous waste inspectors, compliance officers, and permit writers for internal comments.
3. Revise existing state rules. The concept paper and internal comments are reviewed and used to revise the existing state rules. The revisions include changing, adding, and deleting words and languages in the current texts. The revisions may also include the renumbering of the existing rules. Before finalizing the draft rules, it is necessary to circulate the document back to the RCRA compliance enforcement staff for a second-round internal review.
4. Solicit comments from the public. Because the revised rules are subject to public comment, the public should be given the opportunity to review and comment on the revised rules. The public review process is delineated in the state's rulemaking procedure. The procedure specifies, for example, where the draft rules should be announced and by what media, and it describes the frequency of the announcement, the availability of public meetings, and the time frame for public comments.
5. Finalize the proposed rules. Comments from the public are reviewed and used to modify and finalize the proposed rules. Coordination with the state's Attorney General's Office is crucial in this process. An

environmental impact statement should be attached to the proposed rules. The statement describes the impact of the proposed rules to the regulated industry, especially to small businesses. Once completed, the package is ready for submission to the state's regulatory review council for formal processing.

The informal process may take 3 to 6 months depending on the complexity of the proposed rules. Generally, rules containing more complex and controversial issues take longer processing time than those containing relatively simple revisions.

The formal rulemaking process starts with the submission of the proposed rules to the state's regulatory review council and continues with the following steps:

1. The summary of rulemaking and notice of a council meeting. This announcement is published in the state's administrative register.
2. The council conducts a meeting to review and provide comments and decisions on the proposed rules. They may approve, disapprove, or request revisions to the proposed rules.
3. The council's approval and notice of rulemaking. This announcement includes the opportunity for public comment and is published in the state's administrative register.
4. Oral proceedings would be held at the end of the public comment period. This meeting provides an opportunity for the public to deliver oral and written comments on the proposed rules.
5. The adoption of the proposed rules. This final step of the process requires the state's regulatory agency to provide a concise explanatory statement which responds to the public comments before the state adopting the new rules.

The formal process may take 6 to 9 months. As in the informal process, this time frame is also dependent on the nature and complexity of the proposed rules. The above process is a generalized description of the rulemaking procedure taken from Arizona's procedure. Rulemaking procedures vary from state to state.

REGULATORY REFERENCE

The authorization provisions result in grouping states into one of three major categories. A state may have (1) no authorization at all, in which case the federal regulations apply; (2) authorization for part but not all of its program, in which case only those relevant state rules that are part of the authorized portion of the program apply, and federal regulations are otherwise applicable; and (3) authorization for all of its program, in which case state rules apply (Morgulec and Woitte, 1991).[105] In reality, most of the states belong to the second category.

Because of this dual authority situation, it is necessary for hazardous waste planners to contact their state hazardous waste management agency for a copy of the state regulations and information regarding differences between federal and state programs and regulations. Hazardous waste regulatory guidance focusing on state requirements that are more stringent than the federal requirements is also available in most libraries.

More stringent requirements are found, for example, in state provisions dealing with small-quantity generator exclusions, identification and listing, manifests, labeling, transporter permits, liability insurance, groundwater monitoring, required forms, etc. (Keller and Associates, 1993).[106] Because most states adopt the federal regulations by reference, it is normally necessary to obtain a copy of the federal regulations that are referenced by state rules.

Part Three — Data Management

Data management is fundamental to hazardous waste planning implementation. The two most important aspects are the rationale for the data collection and the system approach to data analysis. Chapter 5 discusses this matter by beginning with the question of how a hazardous waste planner recognizes the universe of regulated hazardous wastes a facility must monitor and control. It goes on to describe hazardous waste record-keeping and reporting requirements. Specifically, the chapter discusses hazardous waste manifests, the biennial reporting system, accurate reporting, and the issues related to the administration of hazardous waste reporting requirements.

Chapter 6 examines system approaches to hazardous waste data analysis. The chapter provides a conceptual approach to state hazardous waste flow based on applicable rules and regulations. Issues related to hazardous waste data management are also discussed in this chapter.

5 Hazardous Waste Data Requirements and Management

The hazardous waste management planning process requires the collection of substantial quantities of data. Management, in general, is often defined as a series of processes by which a business or organization sets the common goals and directs its activities toward achieving the goals. Such processes include setting up plans and making various decisions.

In order to accomplish its mission, goals, and objectives, an organization must consider these two processes to be significant. As an integral part of the management process, proper planning and decision-making relies heavily on consistent, timely, and accurate data sets.

Managing hazardous waste data needs some understanding of hazardous waste data sources, the nature of the data, and data analysis techniques. Eventually, a data management system should be able to convert the data into meaningful information. Patton (1988)[107] argues that if data alone could be used to solve planning problems, there would be little reason to convert them into information. Not only must planners understand the information they produce from various data sources, they must also be able to communicate the information precisely to clients, decision-makers, and the public. The process of developing information for planning involves not only locating data, but also evaluating its meaning and its relevance to planning.

There are a number of tasks necessary for planners to develop a basic hazardous waste data management system. The tasks include:

1. identification of hazardous waste data sources,
2. collection and compilation of appropriate data, and
3. analysis of the compiled data and conversion into meaningful hazardous waste information.

While the third task will be discussed in Chapter 6, the first two tasks are covered in the following sections.

This chapter is designed in such a way that, before initiating a discussion on the identification of data sources and the collection and compilation of hazardous waste data, it is appropriate if we first determine whether a facility is indeed handling a hazardous waste. This implies that at the very beginning of the data management process, the hazardous waste planner recognizes the spectrum of the wastes under consideration.

The first step in determining if a facility manages hazardous waste is the application of the regulatory definition of solid waste to the waste streams at the facility. Solid waste need not be in a solid physical state; it may be in a liquid, semi-solid, or even a contained-gaseous physical state.

Regulatory definitions of solid waste and hazardous waste are provided in Chapter 3. The following sections extend the discussion to also include the regulatory exclusion, recycling, and other issues related to solid waste identification and hazardous waste determination.

IDENTIFYING SOLID WASTE

By definition, a solid waste is any material not excluded from regulations which is disposed of, abandoned, burned, recycled, or considered "inherently waste-like." This last phrase refers to dioxin-contaminated waste and furan-contaminated waste.

The two basic sets of full exclusion are wastes excluded from the definition of solid waste, and wastes excluded from the definition of hazardous waste. The subsequent paragraphs discuss the scope of each exclusion.

Under 40 CFR 261.4(a) the following materials are *excluded from the definition of solid waste:*

1. domestic sewage and mixtures of domestic sewage and other wastes passing through a sewer system to a publicly owned treatment works (POTW),
2. industrial point source wastewater discharge requiring a National Pollutant Discharge Elimination System (NPDES) permit,
3. irrigation return flows,
4. radioactive materials defined under the Nuclear Regulatory Commission (NRC),
5. in-situ mining waste not removed from the ground,
6. pulping liquors reclaimed in a pulping liquor recovery furnace and then reused in the pulping process,
7. spent sulfuric acid used to produce virgin sulfuric acid, and
8. secondary materials reclaimed and returned to the original process(es) in which they are generated under certain provisions (specified in 40 CFR 261.4[a]).

A material not considered a solid waste cannot be a hazardous waste. If a material is a solid waste, it might be a hazardous waste. On the other hand, a hazardous waste might be considered nonhazardous based on the regulatory exclusion. 40 CFR 261.4(b) provides such exclusion. Under this provision, the following wastes are *excluded from the definition of hazardous waste:*

1. household hazardous wastes;
2. solid wastes from agricultural crops and animals returned to the soil as fertilizer;
3. mining overburden returned to the mine site;
4. fly ash waste, bottom ash waste, slag waste, and flue gas emission control waste, generated primarily from the combustion of coal or other fossil fuels. This exclusion does not apply to a residue derived from the burning or processing of hazardous waste in a boiler or industrial furnace unless certain specifications are met;
5. drilling fluids, produced waters, and other wastes associated with the exploration, development, or production of crude oil, natural gas, or geothermal energy;
6. wastes containing chromium in the trivalent state and failing the toxicity test solely because of chromium (examples are given in 40 CFR 261.4[b][6][ii]);
7. solid wastes from the extraction, beneficiation, and processing of ores and minerals (with examples and exceptions given in 40 CFR 261.4[b][7]);
8. cement kiln dust waste; and
9. discarded wood or wood products generated by users of arsenical-treated wood or wood products.

In addition to the exclusions described above, if a generator is involved in a recycling activity, the generator must determine if and how it is regulated under hazardous waste regulations. The recycling of hazardous waste presents a unique situation concerning regulatory requirements.

Historically, provided that a generator was going to legitimately recycle a hazardous waste, few regulatory controls were instituted. Because of recent widespread sham recycling, the EPA decided to institute controls to prevent sham recycling while trying to encourage legitimate recycling. The EPA concluded that by redefining solid waste, these goals could be met. The result has been one of the most complicated and confusing set of regulations ever developed by the EPA (Wagner, 1990).[108]

For a generator involved in recycling, Wagner advises that the generator must determine three key factors:

1. What type of recyclable material is the waste?
2. What is the recycling process to be used?
3. Would the material be a hazardous waste if recycling was not pursued?

Each of these factors will assist the generator in determining if the waste, when recycled, is classified as a solid waste.

- If it is not a solid waste, the waste is not regulated as hazardous waste.
- If it is a solid waste, the waste is regulated as hazardous waste.

The following information elaborates on this determination process. Under 40 CFR 261.2, used or spent materials, sludges, and by-products uncovered in a facility are considered recyclable (secondary) materials.

- **Sludges** are residues from treating air, wastewater, or other residues from pollution control.
- **By-products** are residual materials resulting from industrial, commercial, mining, and agricultural operations that are not primary products, are not produced intentionally, and are not fit for a desired end use without further processing.

These *secondary materials* are further classified as follows:

1. spent materials,
2. sludges listed in 40 CFR 261.31 and 261.32,
3. sludges exhibiting characteristics of hazardous waste,
4. by-products listed in 40 CFR 261.31 and 261.32,
5. by-products exhibiting characteristics of hazardous waste,
6. commercial chemical products listed in 40 CFR 261.33, and
7. scrap metal.

Hazardous waste regulations determine that a *recycling activity* can be identified by one of the following components:

1. Used in a manner constituting disposal. This practice involves placing wastes onto the land that contain hazardous wastes as an ingredient. A classic example is that certain wastes are incorporated into asphalt, then used in road construction.
2. Burning for energy recovery. This includes using wastes to produce a fuel.
3. Reclaimed. This refers to the recovery of usable products or regeneration of materials from the wastes.
4. Accumulated speculatively. Speculative accumulation occurs because certain wastes are potentially recyclable, but no feasible recycling market exists for them. It also involves recycling less than 75% of accumulated wastes during a one-year period.

By knowing both the type of secondary material and the recycling process to be used, 40 CFR 261.2 provides a matrix (Table 5.1) showing the relationship between the two elements. Of the 28 combinations available in the matrix, only four combinations (the "NO" cells in Table 5.1) represent recycling exemptions. They are:

1. reclamation of sludges exhibiting characteristics of hazardous waste,
2. reclamation of by-products exhibiting characteristics of hazardous waste,

Table 5.1 Classification of Secondary Materials

Secondary material	Use constituting disposal	Reclamation	Speculative accumulation	Burned as fuel
Spent materials	—	—	—	—
Sludges, listed	—	—	—	—
Sludges, characteristics	—	NO	—	—
By-products, listed	—	—	—	—
By-products, characteristics	—	NO	—	—
Commercial chemical products	—	NO	NO	—
Scrap metal	—	—	—	—

Note: — means the material is a solid waste; "NO" means the material is not a solid waste.

3. reclamation of commercial chemical products, and
4. accumulation of commercial chemical products. Failure to recycle at least 75% of these materials within a year may result in the revocation of this exemption.

Materials described in the above exemptions are not considered solid waste, and therefore, they are not regulated. Once a material is legally classified as a solid waste, the next step is to determine if it is also hazardous.

DETERMINING HAZARDOUS WASTE

Under hazardous waste regulations, a solid waste is a hazardous waste if any of the following conditions apply:

1. It meets the description of any of the lists of hazardous waste to qualify as a listed hazardous waste (either as F-waste, K-waste, P-waste, or U-waste).
2. It is a mixture that contains a listed hazardous waste.
3. It is derived from the treatment, storage, or disposal of a listed hazardous waste.
4. It exhibits any of the ignitability, corrosivity, reactivity, and toxicity characteristics of a hazardous waste.

The description of the listed wastes, the mixture and derived-from rule, and characteristic wastes have been extensively discussed in Chapter 3. Hazardous waste planners should be aware, however, of significant differences between listed wastes and characteristic wastes.

The EPA has specifically listed certain wastes as hazardous. If a generator produces a waste that meets the description contained in the listing, it is presumed to be hazardous regardless of its actual chemical composition. The significance of a waste being listed as hazardous is that a listed waste remains a listed hazardous waste regardless of subsequent treatment or other physical or chemical changes — unless it is delisted through a separate administrative proceeding.

Characteristic wastes are identified as hazardous on the basis of performance tests to discover the presence of specific hazardous characteristics of a waste material. The waste will remain hazardous as long as it continues to exhibit the hazard characteristics. However, a characteristic waste can be rendered nonhazardous through treatment of the waste so that it no longer exhibits the hazardous characteristics.

If a solid waste is a hazardous waste based on one of the above criteria, an equally important determination is to know when the designation of hazardous waste is applied to a material. In general, materials become hazardous waste at the point in time when their intended use has ceased (e.g., obsolete commercial chemical products), or when it is no longer possible to use it (e.g., spent solvent), and the materials are segregated, accumulated, or stored for treatment or disposal, or for reuse or reclamation. Hazardous waste regulations also specify when a hazardous waste ceases to be so defined (40 CFR 261.3[c] and [d]).

The EPA has recognized that a waste from a particular generator may fit the definition of listed hazardous waste but the waste is not actually hazardous. This could be the case if the waste stream does not contain the constituents for which the listed waste was originally identified as hazardous, or if the concentration of the constituents was relatively low.

These situations occur if a facility uses different processes or raw materials than were assumed when the listing was written. Under 40 CFR 260.20 and 260.30, the EPA provides procedures for anyone to petition the agency to delist such waste streams from regulation.

It is common to see businesses handle used industrial oils. In many industrial operations, a large percentage of used industrial oils are recycled by being burned for energy recovery. Fuels derived from used industrial oils are considered effective substitutes for virgin fuels when they have a heat value greater than 5,000 BTUs per pound. However, as explained in the following section, hazardous waste planners should be cautious when dealing with the burning of wastes for energy recovery, which is one of the options among recycling activities.

HAZARDOUS WASTE FUEL AND USED OIL FUEL

Hazardous waste — and mixtures of hazardous waste with used oil — when burned for energy recovery is termed "hazardous waste fuel." Hazardous waste fuel is regulated under 40 CFR Part 266 Subpart D. Part 266 defines standards for the management of special hazardous wastes and specific types of hazardous waste management facilities, and Subpart D is dedicated to hazardous waste burned for energy recovery. However, if hazardous waste generated by a small-quantity generator (SQG) is mixed with used oil (but not with other hazardous wastes), it is not regulated under the hazardous waste fuel regulations but under the used oil regulations (40 CFR 266.40[d][2]).

Hazardous waste fuel regulations prohibit the burning of hazardous waste fuels for energy recovery in anything other than industrial boilers or furnaces that meet certain specifications. The burning of hazardous waste fuels other than for energy recovery is regarded as either treatment or disposal and is subject to treatment, storage, and disposal (TSD) requirements.

Used oil exhibiting a characteristic of hazardous waste without being a listed hazardous waste under the mixture rule or the derived-from rule is termed "used oil fuel" when burned for energy recovery. Used oil fuel is regulated under 40 CFR Part 266, Subpart E: Used Oil Burned for Energy Recovery. As defined in 40 CFR 266.40(b), used oil is any oil that has been refined from crude oil and, as a result of use, is contaminated by physical or chemical impurities. Used oil fuel includes any fuel produced from used oil by processing, blending, or other treatment.

Quality specifications for used oil fuels are found in 40 CFR 266.40(e). The used oil properties and allowable level of each property are specified as follows:

Arsenic	5 ppm	maximum
Cadmium	2 ppm	maximum
Chromium	10 ppm	maximum
Lead	100 ppm	maximum
Flash point	100° F	minimum
Total halogens	4000 ppm	maximum

When used oil is burned for energy recovery and exceeds any of the above specification levels, the used oil is subject to 40 CFR Part 266 Subpart E. This type of used oil is termed *off-specification used oil.* Off-specification used oil can only be burned in industrial boilers and furnaces as specified in 40 CFR 266.41(b).

It is the responsibility of a hazardous waste planner to determine if the facility handles hazardous waste and to know what types of waste are handled, how much waste is generated on a monthly basis, and how much is shipped off-site from the facility. Hazardous waste regulations establish record-keeping and reporting requirements for hazardous waste handlers and provide the handlers with record-keeping and reporting compliance guidance.

For a state, the requirements represent a mechanism to track the quantities of waste managed and transported by hazardous waste handlers statewide. These record-keeping and reporting provisions consist of the hazardous waste manifest requirement and the submission of the biennial report by hazardous waste handlers.

HAZARDOUS WASTE MANIFEST

As codified in the hazardous waste regulations, any person who has hazardous waste transported off-site must prepare a manifest. The manifest must be signed and also include the handwritten signature of the initial transporter and the date of acceptance. The waste shipment from a point of origin to its destination may utilize one or more transporters.

The generator must retain one copy and give the transporter the remaining copies of the manifest. Copies of each manifest must be kept for three years by the generator after they are signed by the designated facility that received the waste and returned to the generator. The 3-year period is determined from the date the waste was accepted by the initial transporter.

The manifest requirement is applicable to large-quantity generators (LQGs) and small-quantity generators (SQGs). An LQG is any facility owner/operator, or person who generates more than 2,200 lb of hazardous waste or 2.2 lb of acutely hazardous waste in any calendar month. A generator of more than 220 lb but less than 2,200 lb of hazardous waste or less than 2.2 lb of acutely hazardous waste in any month is defined as an SQG. If a generator produces not more than 220 lb of hazardous waste in any calendar month, it is a conditionally exempt SQG and not subject to the manifest requirement. The manifest requirement is not applicable to used oil; however, some states may require manifesting of used oil shipments.

A manifest is designed so that generators can track the movement of hazardous waste from the point of generation to the point of ultimate destination. The manifest is part of a controlled tracking system. Each time the waste is transported off-site, the manifest must be signed to acknowledge receipt of the waste. Once the waste is received by the designated facility, the owner or operator of that facility must send a copy of the manifest back to the generator. This system ensures that the generator has documentation that the hazardous waste has made it to its ultimate destination. Because the purpose of a manifest is to track hazardous waste shipped off-site, it is not required of generators who treat, store, or dispose of their waste on-site.

The hazardous waste manifest serves as both a shipping document and a tracking document. The manifest provides important information including:

1. the name and location of the generator and the receiving facility, as well as the identity of the transporter(s);
2. the description and the waste code of the waste being transported; and
3. the quantities of the waste being transported.

Figure 5.1 shows a copy of the EPA's Uniform Hazardous Waste Manifest form. The form has been designed for use on a 12-pitch (elite) typewriter, but pen may also be used by pressing firmly. A total of 19 features are itemized on the form. Items 11, 12, 15, and 16 are briefly discussed in the next paragraphs.

Item 11 requires the generator to enter, for each hazardous waste shipped, the U.S. Department of Transportation (DOT) proper shipping name, hazard class, and United Nations (UN) or North America (NA) identification number of the waste. Examples are:

Shipping name	Hazard class	ID number
Acetone	Flammable liquid	UN 1090
Chloric acid	Oxidizer	NA 2626
Methyl ethyl ketone	Flammable liquid	UN 1193

Section 172.101 of 49 CFR provides a table containing the above information including the packaging and labeling requirements. The UN identification number is used for international as well as domestic shipments. The NA identification number is not recognized for international shipments and can only be used in North America. Hazard classes include the following traits: combustible, corrosive, explosive, flammable, oxidizer, poisonous, and radioactive. Please note that Title 49 of the CFR deals with the transportation of hazardous materials. Under DOT regulations, hazardous waste is considered a hazardous material for the purpose of shipment.

The hauler of the waste or the designated TSD facility may be able to help determine the shipping requirements. Assistance can also be obtained from the DOT at (202) 366-5580, the state's department of transportation, the state's environmental agency, or trade associations.

Item 12 requires the generator to enter the number of containers for each waste shipped and the appropriate abbreviation for the type of container used for the shipment. The container coding is *listed on the back* of the manifest. Examples of the abbreviations used are:

DM = metal drums, barrels, kegs
DW = wooden drums, barrels, kegs
TP = portable tanks

Generators may use **item 15** to indicate special handling instructions, or transportation, treatment, storage, or disposal information, or Bill of Lading information. A generator must certify in **item 16** that he has made a good-faith effort to minimize his waste generation and to select the best waste management method that he can afford.

It is extremely important for the generator to prepare the manifest properly, because the generator is responsible for the hazardous waste he produces and its ultimate disposition. A good manifest data system is vital to any facility. It is used for regulatory compliance and also represents an essential tool of the facility's hazardous waste management planning process.

A good manifest data system enables hazardous waste planners to retrieve the facility's basic hazardous waste information on a monthly, quarterly, or yearly basis. Examples of the types of information that can be retrieved from the manifest data system are:

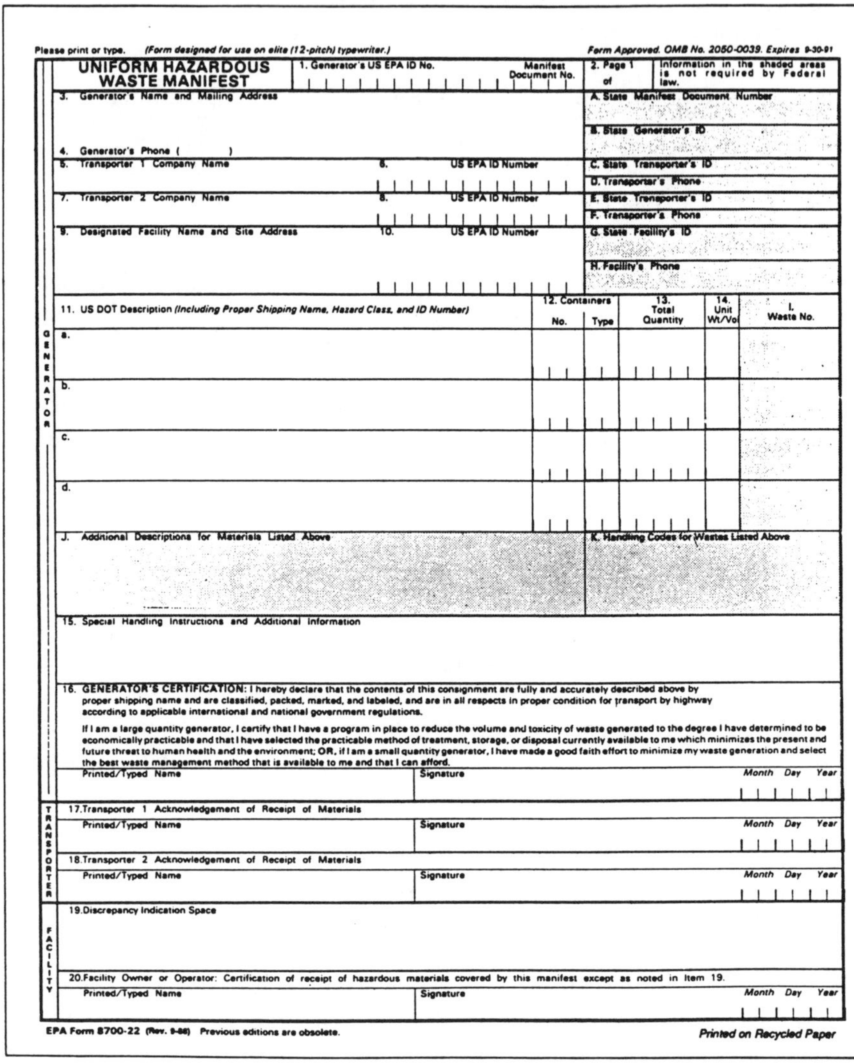

Please print or type. *(Form designed for use on elite (12-pitch) typewriter.)* *Form Approved. OMB No. 2050-0039. Expires 9-30-91*

UNIFORM HAZARDOUS WASTE MANIFEST

1. Generator's US EPA ID No.

Manifest Document No.

2. Page 1 of

Information in the shaded areas is not required by Federal law.

3. Generator's Name and Mailing Address

A. State Manifest Document Number

B. State Generator's ID

4. Generator's Phone ( )

5. Transporter 1 Company Name

6. US EPA ID Number

C. State Transporter's ID

D. Transporter's Phone

7. Transporter 2 Company Name

8. US EPA ID Number

E. State Transporter's ID

F. Transporter's Phone

9. Designated Facility Name and Site Address

10. US EPA ID Number

G. State Facility's ID

H. Facility's Phone

GENERATOR

11. US DOT Description *(Including Proper Shipping Name, Hazard Class, and ID Number)*

12. Containers No. Type

13. Total Quantity

14. Unit Wt/Vol

I. Waste No.

a.

b.

c.

d.

J. Additional Descriptions for Materials Listed Above

K. Handling Codes for Wastes Listed Above

15. Special Handling Instructions and Additional Information

16. GENERATOR'S CERTIFICATION: I hereby declare that the contents of this consignment are fully and accurately described above by proper shipping name and are classified, packed, marked, and labeled, and are in all respects in proper condition for transport by highway according to applicable international and national government regulations.

If I am a large quantity generator, I certify that I have a program in place to reduce the volume and toxicity of waste generated to the degree I have determined to be economically practicable and that I have selected the practicable method of treatment, storage, or disposal currently available to me which minimizes the present and future threat to human health and the environment; OR, if I am a small quantity generator, I have made a good faith effort to minimize my waste generation and select the best waste management method that is available to me and that I can afford.

Printed/Typed Name | Signature | *Month Day Year*

TRANSPORTER

17. Transporter 1 Acknowledgement of Receipt of Materials

Printed/Typed Name | Signature | *Month Day Year*

18. Transporter 2 Acknowledgement of Receipt of Materials

Printed/Typed Name | Signature | *Month Day Year*

FACILITY

19. Discrepancy Indication Space

20. Facility Owner or Operator: Certification of receipt of hazardous materials covered by this manifest except as noted in Item 19.

Printed/Typed Name | Signature | *Month Day Year*

EPA Form 8700-22 (Rev. 9-88) Previous editions are obsolete. *Printed on Recycled Paper*

Figure 5.1 Uniform Hazardous Waste Manifest.

1. the amount of hazardous waste shipped off-site by the facility, by type of hazardous waste;
2. the amount shipped by names of the receiving facility; and
3. the amount shipped by names of the transporter.

At the state level, the use of the manifest data in the state's hazardous waste management planning is somewhat restricted. Limitations embodied in the manifest requirements are:

Please print or type. (Form designed for use on elite (12-pitch) typewriter.) Form Approved. OMB No. 2050-0039. Expires 9-30-91

UNIFORM HAZARDOUS WASTE MANIFEST *(Continuation Sheet)*

21. Generator's US EPA ID No. | Manifest Document No. | 22. Page | Information in the shaded areas is not required by Federal law.

23. Generator's Name | L. State Manifest Document Number | M. State Generator's ID

24. Transporter ____ Company Name | 25. US EPA ID Number | N. State Transporter's ID | O. Transporter's Phone

26. Transporter ____ Company Name | 27. US EPA ID Number | P. State Transporter's ID | Q. Transporter's Phone

28. US DOT Description *(Including Proper Shipping Name, Hazard Class, and ID Number)*	29. Containers No.	29. Containers Type	30. Total Quantity	31. Unit Wt/Vol	R. Waste No.
a.					
b.					
c.					
d.					
e.					
f.					
g.					
h.					
i.					

S. Additional Descriptions for Materials Listed Above | T. Handling Codes for Wastes Listed Above

32. Special Handling Instructions and Additional Information

33. Transporter ____ Acknowledgement of Receipt of Materials | Date
Printed/Typed Name | Signature | Month Day Year

34. Transporter ____ Acknowledgement of Receipt of Materials | Date
Printed/Typed Name | Signature | Month Day Year

35. Discrepancy Indication Space

GENERATOR / TRANSPORTER / FACILITY

EPA Form 8700-22A (Rev. 9-88) Previous edition is obsolete. *Printed on Recycled Paper*

Figure 5.1 (Continued.)

1. A generator is not required to submit a copy of the manifest to the state, although some states have mandated generators to send copies of the signed manifests to the state's environmental agency.
2. Conditionally exempt SQGs are exempted from the manifesting requirement. This means that the manifest data system only provides information from generators who produce more than 220 lb of hazardous waste in any calendar month.

3. The manifest deals with off-site shipments only. It does not take into account waste generated and treated, stored, or disposed on-site. As such, the manifest does not account for the waste generated in-state that is not shipped off-site.

In addition to the manifest requirement, a generator who handles hazardous waste in a regulated quantity must also prepare and submit a copy of a biennial report to the EPA, or the authorized state, by March 1 of each even-numbered year. The report must be submitted on an EPA-approved format and must cover the handler's activities during the previous year. The reporting forms and accompanying instructions are collated in a bound document popularly known as the biennial reporting system (BRS) booklet. The document is available at the state's environmental agency.

BIENNIAL REPORTING SYSTEM

The biennial reporting system perhaps represents the most important data-management tool for hazardous waste planners. In this reporting mechanism, the generators are required to specify the characteristics of the waste generated in the reporting year and the activities conducted to manage the waste, such as storage, recycling, treatment, transportation, or disposal. The system was designed by the EPA to achieve a number of national objectives. The objectives are:

1. to provide the EPA and the state with an understanding of hazardous waste generation, management, and waste minimization activities in the state;
2. to help measure the quality of the environment;
3. to assist the state in preparing a hazardous waste Capacity Assurance Plan (CAP) as required by the Comprehensive Environmental Response, Compensation, and Liability Act of 1980 (CERCLA) (the CAP process will be discussed in Chapter 7); and
4. to communicate the findings to the public primarily through the publication of the National BRS Report (U.S. EPA, 1993).[109]

Structurally, the most recent (1993) biennial report consists of five forms. They are:

1. identification and certification (IC) form,
2. waste generation and management (GM) form,
3. waste received from off-site (WR) form,
4. waste treatment, disposal, or recycling process system (PS) form, and
5. off-site identification (OI) form.

Each of these forms and the procedures to complete the forms properly are scribed in the next five sections. The EPA also provides the BRS help line. To tain assistance in completing the forms, you may call (800) 908-2159. The help line only operates from January through April of even-numbered years, and the toll-free number may change for the next reporting period.

IC Form

The identification and certification (IC) form (Figure 5.2) requires that a facility provide its facility name, EPA ID number, address, name and phone number of the facility's contact person, and the standard industrial classification (SIC) codes that describe the principal products, group of products produced or distributed, or the service rendered at the site's location.

SIC codes are organized by industrial activity. Examples of such codes include SIC code 2711 for newspaper publishing, 2911 for petroleum refinery, 3366 for copper foundries, 3672 for printed circuit board manufacturers, 4213 for trucking, and 7537 for auto repair shops. A complete list of SIC codes is provided in the BRS booklet.

Figure 5.2 displays a copy of the IC form. The figure shows that the identification and certification form also requires the generator to record information on the waste minimization program undertaken by the facility. This part is easy because the generator is only required to make check marks on a highly structured questionnaire.

The BRS booklet defines waste minimization as the reduction — to the extent feasible — of waste that is generated or subsequently treated, stored, or disposed. Waste minimization includes:

1. source reduction, and/or
2. recycling activity,

undertaken by a generator, that results in the reduction of the total quantity of hazardous waste, the reduction of toxicity of the hazardous waste, or both, as long as the reduction is consistent with the goal of minimizing present and future threats to human health and the environment.

Source reduction means the reduction of hazardous waste at the source, usually within a process. The term implies any action that reduces the quantity or the toxicity of the waste exiting a process.

Recycling means:

1. the use or reuse of waste as an effective substitute for a commercial product or as an ingredient or feedstock in an industrial process, and
2. the reclamation of useful constituent fractions within a waste material or the removal of contaminants from a waste to allow it to be reused. The term implies any action undertaken after the waste has been generated.

The certification box on the form must be signed by the generator. By signing, the generator certifies that the information reported throughout is truthful, accurate, and complete.

BEFORE COPYING FORM, ATTACH SITE IDENTIFICATION LABEL OR ENTER:

SITE NAME:

EPA ID NO:

U.S. ENVIRONMENTAL PROTECTION AGENCY

1993 Hazardous Waste Report

FORM IC

IDENTIFICATION AND CERTIFICATION

INSTRUCTIONS: Read the detailed instructions beginning on page 9 of the 1993 Hazardous Waste Report booklet before completing this form.

Sec. I Site name and location address. Complete A through H. Check the box ☐ in items A, C, E, F, G, and H if same as label; if different, enter corrections. If label is absent, enter information. Instruction page 10.

A. EPA ID No. Same as label ☐ or →	B. County
C. Site/company name Same as label ☐ or →	D. Has the site name associated with this EPA ID changed since 1991? ☐ 1 Yes ☐ 2 No

E. Street name and number. If not applicable, enter industrial park, building name, or other physical location description.
Same as label ☐ or →

F. City, town, village, etc. Same as label ☐ or →	G. State Same as label	H. Zip Code Same as label

Sec. II Mailing address of site. Instruction page 10.

A. Is the mailing address the same as the location address? ☐ 1 Yes (SKIP TO SEC. III) ☐ 2 No (GO TO BOX B)

B. Number and street name of mailing address

C. City, town, village, etc.	D. State	E. Zip Code

Sec. III Name, title, and telephone number of the person who should be contacted if questions arise regarding this report. Instruction page 10.

A. Please print: Last Name First name M.I.	B. Title	C. Telephone Extension

Sec. IV "I certify under penalty of law that this document and all attachments were prepared under my direction or supervision in accordance with a system designed to assure that qualified personnel properly gather and evaluate the information submitted. Based on my inquiry of the person or persons who manage the system, or those persons directly responsible for gathering the information, the information submitted is, to the best of my knowledge and belief, true, accurate and complete. I am aware that there are significant penalties under Section 3008 of the Resource Conservation and Recovery Act for submitting false information, including the possibility of fine and imprisonment for knowing violations."

A. Please print: Last Name First name M.I.	B. Title
C. Signature	D. Date of signature MO. DAY YR.

Figure 5.2 Identification and Certification (IC) Form.

GM Form

This second form is divided into four sections that together document the source, the characteristics, the amount of hazardous waste generated and handled on-site, and the handling method. Also provided in the waste generation and management (GM) form (Figure 5.3) is the quantity of hazardous waste shipped off-site, the off-site handling methods, and the waste minimization activities related to the hazardous waste stream. A GM form must be completed for each hazardous waste stream generated.

Sec.V - Generator Status

EPA ID NO. |_|_|_|_| |_|_|_|_| |_|_|_|_| |_|_|_|_|

A. 1993 RCRA generator status Instruction page 10. (CHECK ONE BOX BELOW)	B. Reason for not generating Page 12. (CHECK ALL THAT APPLY)	
□ 1 LQG □ 2 SQG — SKIP to SEC. VI □ 3 CESQG □ 4 Non generator (Continue to Box B)	□ 1 Never generated □ 2 Out of business □ 3 Only excluded or delisted waste □ 4 Only non-hazardous waste	□ 5 Periodic or occasional generator □ 6 Waste minimization activity □ 7 Other (SPECIFY COMMENTS IN BOX BELOW)

Sec.VI - On-Site Waste Management Status

A. Storage subject to RCRA permitting requirements Page 13.	B. Treatment, disposal, or recycling subject to RCRA permitting requirements Page 13.	C. RCRA-exempt treatment, disposal, or recycling Page 13.
|_|	|_|	|_|

Sec.VII - Waste Minimization Activity during 1992 or 1993

A. Did this site begin or expand a source reduction activity during 1992 or 1993? Page 14.	B. Did this site begin or expand a recycling activity during 1992 or 1993? Page 15.	C. Did this site systematically investigate opportunities for source reduction or recycling during 1992 or 1993? Page 15.
□ 1 Yes □ 2 No	□ 1 Yes □ 2 No	□ 1 Yes □ 2 No

D. Did any of the factors listed below delay or limit this site's ability to initiate new or additional source reduction activities in 1992 or 1993? Page 15
(CHECK YES OR NO FOR EACH ITEM)

Yes	No		
□ 1	□ 2	a.	Insufficient capital to install new source reduction equipment or implement new source reduction practices
□ 1	□ 2	b.	Lack of technical information on source reduction techniques applicable to the specific production processes
□ 1	□ 2	c.	Source reduction is not economically feasible: cost savings in waste management or production will not recover the capital investment
□ 1	□ 2	d.	Concern that product quality may decline as a result of source reduction
□ 1	□ 2	e.	Technical limitations of the production processes
□ 1	□ 2	f.	Permitting burdens
□ 1	□ 2	g.	Source reduction previously implemented - additional reduction does not appear to be technically feasible
□ 1	□ 2	h.	Source reduction previously implemented - additional reduction does not appear to be economically feasible
□ 1	□ 2	i.	Source reduction previously implemented - additional reduction does not appear to be feasible due to permitting requirements
□ 1	□ 2	j.	Other (SPECIFY COMMENTS IN BOX BELOW)

E. Did any of the factors listed below delay or limit the site's ability to initiate new or additional on-site or off-site recycling activities during 1992 or 1993? Page 15.
(CHECK YES OR NO FOR EACH ITEM)

Yes	No		
□ 1	□ 2	a.	Insufficient capital to install new recycling equipment or implement new recycling practice
□ 1	□ 2	b.	Lack of technical information on recycling techniques applicable to this site's specific production process
□ 1	□ 2	c.	Recycling is not economically feasible: cost savings in waste management will not recover the capital investment
□ 1	□ 2	d.	Concern that product quality may decline as a result of recycling
□ 1	□ 2	e.	Requirements to manifest wastes inhibit shipments of off-site for recycling
□ 1	□ 2	f.	Financial liability provisions inhibit shipments off-site for recycling
□ 1	□ 2	g.	Technical limitations of production processes inhibit shipments off-site for recycling
□ 1	□ 2	h.	Technical limitations of production processes inhibit on-site recycling
□ 1	□ 2	i.	Permitting burdens inhibit recycling
□ 1	□ 2	j.	Lack of permitted off-site recycling facilities
□ 1	□ 2	k.	Unable to identify a market for recycled materials
□ 1	□ 2	l.	Recycling previously implemented - additional recycling does not appear to be technically feasible
□ 1	□ 2	m.	Recycling previously implemented - additional recycling does not appear to be economically feasible
□ 1	□ 2	n.	Recycling previously implemented - additional recycling does not appear to be feasible due to permitting requirements
□ 1	□ 2	o.	Other (SPECIFY COMMENTS IN BOX BELOW)

Comments:

Figure 5.2 (Continued.)

Section I of the form requests information on each hazardous waste generated, treated, recycled, or disposed on-site, or shipped off-site during the previous calendar year. Each hazardous waste is identified by using an EPA waste code, such as D002 for corrosive waste, F007 for spent cyanide plating bath solutions from electroplating operations, K052 for tank bottoms (leaded) from the petroleum refining industry, P004 for aldrin, and U247 for methoxychlor. A complete list of EPA waste codes is provided in the BRS booklet. A copy of the **EPA waste code** list is in Appendix V.1.

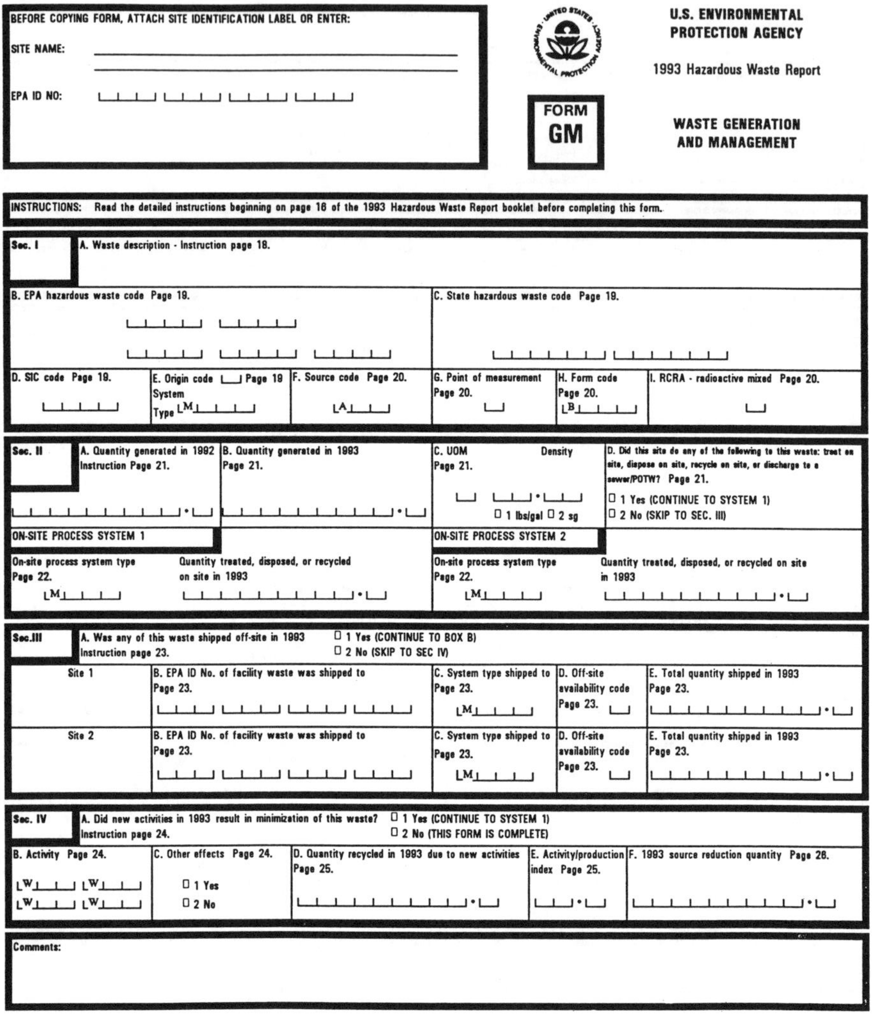

BEFORE COPYING FORM, ATTACH SITE IDENTIFICATION LABEL OR ENTER:

SITE NAME:

EPA ID NO:

U.S. ENVIRONMENTAL PROTECTION AGENCY

1993 Hazardous Waste Report

FORM GM

WASTE GENERATION AND MANAGEMENT

INSTRUCTIONS: Read the detailed instructions beginning on page 16 of the 1993 Hazardous Waste Report booklet before completing this form.

Sec. I — A. Waste description - Instruction page 18.

B. EPA hazardous waste code Page 19. | C. State hazardous waste code Page 19.

D. SIC code Page 19. | E. Origin code Page 19; System Type M | F. Source code Page 20. A | G. Point of measurement Page 20. | H. Form code Page 20. B | I. RCRA - radioactive mixed Page 20.

Sec. II — A. Quantity generated in 1992 Instruction Page 21. | B. Quantity generated in 1993 Page 21. | C. UOM Page 21. Density; □ 1 lbs/gal □ 2 sg | D. Did this site do any of the following to this waste: treat on site, dispose on site, recycle on site, or discharge to a sewer/POTW? Page 21. □ 1 Yes (CONTINUE TO SYSTEM 1) □ 2 No (SKIP TO SEC. III)

ON-SITE PROCESS SYSTEM 1 — On-site process system type Page 22. M; Quantity treated, disposed, or recycled on site in 1993

ON-SITE PROCESS SYSTEM 2 — On-site process system type Page 22. M; Quantity treated, disposed, or recycled on site in 1993

Sec. III — A. Was any of this waste shipped off-site in 1993 Instruction page 23. □ 1 Yes (CONTINUE TO BOX B) □ 2 No (SKIP TO SEC IV)

Site 1 — B. EPA ID No. of facility waste was shipped to Page 23. | C. System type shipped to Page 23. M | D. Off-site availability code Page 23. | E. Total quantity shipped in 1993 Page 23.

Site 2 — B. EPA ID No. of facility waste was shipped to Page 23. | C. System type shipped to Page 23. M | D. Off-site availability code Page 23. | E. Total quantity shipped in 1993 Page 23.

Sec. IV — A. Did new activities in 1993 result in minimization of this waste? Instruction page 24. □ 1 Yes (CONTINUE TO SYSTEM 1) □ 2 No (THIS FORM IS COMPLETE)

B. Activity Page 24. W W W W | C. Other effects Page 24. □ 1 Yes □ 2 No | D. Quantity recycled in 1993 due to new activities Page 25. | E. Activity/production index Page 25. | F. 1993 source reduction quantity Page 26.

Comments:

Figure 5.3 Generation and Management (GM) Form.

The required information associated with the generation of waste includes the SIC code, the source code, the form code, and the system type code. Examples of coding for sources of waste generation are A05 for dip rinsing, A22 for electroplating, A39 for slag removal, A54 for oil changes, and A76 for sludge dewatering. The listing of **source codes** can be found in Appendix V.2.

The forms of waste are similarly coded. For example, B117 represents waste liquid mercury, B212 is reactive or polymerizable organic liquid, B301 is soil contaminated with organics, B302 is soil contaminated with inorganics, B508 is sludge with reactive sulfides, and B603 is oily sludge. The **form codes** are listed in Appendix V.3.

The system type codes start with letter "M." Examples are: M013 for secondary smelting, M023 for solvent extraction, M041 for liquid incineration, M061 for fuel blending, M077 for chemical precipitation, M082 for carbon adsorption, M121 for neutralization, and M132 for landfill. Appendix V.4 provides a list of **system type codes**.

A complete list of source codes, form codes, and system type codes is also provided in the BRS booklet. The above examples indicate that the source code describes the production, service, or waste management process that was the source associated with the generation of the waste. The form code corresponds to the physical and chemical state of the reported hazardous waste. The system type code explains the operation from which the waste is a residual.

Section II of the waste generation and management form requests information on the quantities of hazardous waste generated during the reporting calendar year and the prior year. *Section III* requests information on off-site shipments of hazardous waste. This includes the EPA ID number of the facility to which the waste was shipped, the system type of the receiving facility, and the quantity of the waste shipped.

Section IV, the last section of the GM form, requests information on any new activities undertaken during the reporting year that resulted in waste minimization. The waste minimization activity codes are provided in the BRS booklet. Examples of the coding system are: W34 means secondary containment installed to prevent the generation of hazardous waste due to chemical spills, W42 represents raw material substitution for the purpose of reducing the generation of hazardous waste, and W65 indicates redesigned parts racks to reduce dragout. The waste minimization **activity codes** are provided in Appendix V.5.

The result of waste minimization is indicated by both the quantity of waste recycled and the source reduction quantity during the reporting year. For calibration, the waste generation and management form provides a measure of changes in economic and other factors that affect the comparison of quantities of hazardous waste generated in the observed years. The measure, with examples and calculations, is discussed in the BRS booklet.

WR Form

The waste received (WR) form (Figure 5.4) must be completed only if the facility received RCRA hazardous waste from off-site during the reporting year. For each type of hazardous waste received, the facility must report the quantities and characteristics of the waste in terms of the EPA hazardous waste code, the waste form code, and the system type code. The facility is also required to describe the on-site treatment or disposal system in which the waste was managed.

PS Form

The process system (PS) form (Figure 5.5) must be completed by a TSD facility or generator that treats hazardous waste on-site. The form is divided into two sections that together document the type, utilization, and capacity of a treatment or disposal system at the facility.

BEFORE COPYING FORM, ATTACH SITE IDENTIFICATION LABEL OR ENTER:

SITE NAME: ______________________

EPA ID NO: |_|_|_|_| |_|_|_|_| |_|_|_|_| |_|_|_|_|

FORM **WR**

U.S. ENVIRONMENTAL PROTECTION AGENCY

1993 Hazardous Waste Report

WASTE RECEIVED FROM OFF-SITE

INSTRUCTIONS: Read the detailed instructions beginning on page 30 of the 1993 Hazardous Waste Report booklet before completing this form.

Waste 1	A. Description of hazardous waste Instruction page 30.	B. EPA hazardous waste code Page 31.	C. State hazardous waste code Page 31.
D. Off-site source EPA ID number Page 31.	E. Quantity received in 1993 Page 31.	F. UOM Page 31.	Density ☐ 1 lbs/gal ☐ 2 sg
G. Waste form code Page 32. B	H. RCRA-radioactive mixed Page 32.	I. System type Page 32. M	

Waste 2	A. Description of hazardous waste Instruction page 30.	B. EPA hazardous waste code Page 31.	C. State hazardous waste code Page 31.
D. Off-site source EPA ID number Page 31.	E. Quantity received in 1993 Page 31.	F. UOM Page 31.	Density ☐ 1 lbs/gal ☐ 2 sg
G. Waste form code Page 32. B	H. RCRA-radioactive mixed Page 32.	I. System type Page 32. M	

Waste 3	A. Description of hazardous waste Instruction page 30.	B. EPA hazardous waste code Page 31.	C. State hazardous waste code Page 31.
D. Off-site source EPA ID number Page 31.	E. Quantity received in 1993 Page 31.	F. UOM Page 31.	Density ☐ 1 lbs/gal ☐ 2 sg
G. Waste form code Page 32. B	H. RCRA-radioactive mixed Page 32.	I. System type Page 32. M	

Comments:

Figure 5.4 Waste Received (WR) Form.

The figure shows that *Section I* requests descriptive information on the process, unit type, operational status, and regulatory status of the system. The unit types include tank; container; incinerator; industrial kiln, furnace, or boiler; waste pile; underground injection well; landfill; land treatment; surface impoundment; or another type of unit which is specified in the comments box.

Operational status refers to whether the facility is still planned, under construction, permanently closed, temporarily idle, or operational. Regulatory status

BEFORE COPYING FORM, ATTACH SITE IDENTIFICATION LABEL OR ENTER:

SITE NAME: ____________________

EPA ID NO: |_|_|_|_| |_|_|_|_| |_|_|_|_| |_|_|_|_|

FORM **PS**

U.S. ENVIRONMENTAL PROTECTION AGENCY

1993 Hazardous Waste Report

WASTE TREATMENT, DISPOSAL, OR RECYCLING PROCESS SYSTEMS

INSTRUCTIONS: Read the detailed instructions beginning on page 33 of the 1993 Hazardous Waste Report booklet before completing this form.

Sec. I A. Waste treatment, disposal, or recycling system description
Instruction Page 38.

B. System type Page 38.	C. Regulatory status Page 39.	D. Operational status Page 39.	E. Unit types Page 39.
M____	__	__	__ __

Sec. II A. 1993 influent quantity
Instruction page 40.

Total ____________ • __ UOM __ Density __ • __
RCRA ____________ • __ ☐ 1 lbs/gal ☐ 2 sg

B. Maximum operational capacity
Page 41.

Total ____________ • __
RCRA ____________ • __

C. 1993 liquid effluent quantity
Instruction page 42.

Total ____________ • __ UOM __ Density __ • __
RCRA ____________ • __ ☐ 1 lbs/gal ☐ 2 sg

D. 1993 solid/sludge residual quantity
Page 42.

Total ____________ • __ UOM __ Density __ • __
RCRA ____________ • __ ☐ 1 lbs/gal ☐ 2 sg

E. Limitation on maximum operational capacity Page 43.	F. Commercial capacity availability code Page 43.	G. Percent capacity commercially available Page 43.
1. __ 2. __ 3. __	__	___ %

Comments:

Figure 5.5 Process Systems (PS) Form.

indicates the facility as RCRA regulated, RCRA exempted, state regulated, and other regulatory variations specified in the booklet.

Section II requests information on the quantity of influents, effluents, and residuals associated with the treatment or disposal system that operated at any time during the reporting year. Also requested are the maximum operational capacity of the system and whether the system is available for other companies as a commercial hazardous waste management facility. A **commercial facility** is a facility that receives revenue for providing waste storage, treatment, or disposal service to hazardous waste generators.

BEFORE COPYING FORM, ATTACH SITE IDENTIFICATION LABEL OR ENTER:

SITE NAME: ______________________

EPA ID NO: |_|_|_|_|_|_|_|_|_|_|_|_|

U.S. ENVIRONMENTAL PROTECTION AGENCY

1993 Hazardous Waste Report

FORM OI

OFF-SITE IDENTIFICATION

INSTRUCTIONS: Read the detailed instructions on the reverse side before completing this form.

Site 1	A. EPA ID No. of off-site installation or transporter	B. Name of off-site installation or transporter
C. Handler type	(CHECK ALL THAT APPLY) □ Generator □ Transporter □ TSDR	D. Address of generator Street City State Zip

Site 2	A. EPA ID No. of off-site installation or transporter	B. Name of off-site installation or transporter
C. Handler type	(CHECK ALL THAT APPLY) □ Generator □ Transporter □ TSDR	D. Address of generator Street City State Zip

Site 3	A. EPA ID No. of off-site installation or transporter	B. Name of off-site installation or transporter
C. Handler type	(CHECK ALL THAT APPLY) □ Generator □ Transporter □ TSDR	D. Address of generator Street City State Zip

Site 4	A. EPA ID No. of off-site installation or transporter	B. Name of off-site installation or transporter
C. Handler type	(CHECK ALL THAT APPLY) □ Generator □ Transporter □ TSDR	D. Address of generator Street City State Zip

Site 5	A. EPA ID No. of off-site installation or transporter	B. Name of off-site installation or transporter
C. Handler type	(CHECK ALL THAT APPLY) □ Generator □ Transporter □ TSDR	D. Address of generator Street City State Zip

Comments:

Figure 5.6 Off-Site Identification (OI) Form.

OI Form

The off-site identification (OI) form (Figure 5.6) must be completed only if your state requires it. The purpose of this form is to document the names and addresses of off-site TSD facilities to which a generator shipped his hazardous waste and the transporters used for the shipments during the reporting period.

Completing the BRS forms, hazardous waste manifests, and other record-keeping and reporting documents correctly is critical. Not only is accurate reporting essential for hazardous waste planning, but it is also important for regulatory compliance. Failure to do so may result in unnecessary litigation and bad publicity. Among other provisions, RCRA provides for felony penalties for anyone who knowingly omits material information from — or makes a material false statement or representation in — documents required for purposes of compliance with RCRA Subtitle C (Morgulec and Woitte, 1991).[110]

Accurate reporting is one of the most important issues in hazardous waste data management. A number of issues related to reporting, such as mixed hazardous waste, conversion coefficient, waste coding, and manifest tracking are also important for achieving accurate and reliable hazardous waste data.

ISSUES IN HAZARDOUS WASTE REPORTING

There are two components in the biennial reporting system that represent important issues for hazardous waste planning. One of them is the **mixed hazardous waste** issue. If a generator produced a mixed hazardous waste that consists of a number of waste codes, the EPA allows the generator to allocate equally the quantity of the mixed waste among the waste codes that make up the mixed waste. For example, a mixed hazardous waste of 9 tons consists of F001, F002, and D001. The waste generation should be reported as 3 tons of F001, 3 tons of F002, and 3 tons of D001. This procedure is used simply for practicality. It does not necessarily reflect the accuracy of the reported quantities.

The second issue deals with the **conversion coefficient,** which is used to convert volume to weight. This can be done easily if we know the density of the waste. In most cases, the density is not known specifically, because the reported hazardous waste is thoroughly mixed with various solid waste and/or hazardous waste types. This problem can be addressed by using a "normalized" density, such as using a coefficient of 8.3 lb/gal, or any other number that the planner feels to be the best estimated density for the waste. Obviously, the normalized density represents some form of approximation.

In addition to the above specific issues, the use of the RCRA waste codes for the biennial reporting system (BRS) may also produce inaccurate counts of the quantities of regulated waste streams. This can happen simply due to *misclassification* of the waste (U.S. GAO, 1990).[111] Although classifying a waste into an RCRA waste code sounds simple enough, hazardous wastes defy easy definition, and it is generally difficult to classify waste materials into a package that follows set rules (Cheremisinoff, 1994).[112]

In certain shipping situations, manifest regulations fail to provide procedures that would ensure proper hazardous waste tracking. Marvin (1993)[113] observes a number of situations that represent important planning issues in manifest tracking. He argues that manifests generally are inadequate for tracking hazardous waste when *part of a load is rejected* by a TSD facility. In such instances, the manifest will remain with the accepted waste at the facility. A copy of the original manifest stating the quantity of rejected waste will be given to the transporter as a receipt from the facility. This is not intended as a shipping or tracking document for the rejected manifest.

Marvin asserts that manifests are ill-suited to situations in which wastes are *repackaged before arriving* at a TSD facility. Such repackaging ordinarily occurs at a terminal or transfer facility operated by a transporter when, for example, the wastes are transferred from drums to larger containers. No clear method is specified for indicating on a manifest such repackaging details as date and location.

An additional waste tracking problem may arise, Marvin noted, if waste is shipped in a bulk container such as tank truck or railway tank car to a TSD facility. Because complete removal of bulk waste may be impossible, a *tank with residue* of the original load must be shipped back to the generator or a cleaning facility. Additional documentation is not currently in place to address this situation.

A simple common error in the manifest that cannot be easily detected is applying an *incorrect unit of measure* for the load. Instead of a pound, a generator may write a ton in the manifest — which obviously will create difficulty later in the data processing.

Currently, the manifest is gaining popularity for managing state hazardous waste programs. According to a document released by the National Governors' Association (NGA) in 1991, about two-thirds of the states in the U.S. have established programs to compile manifest records in a computerized manifest data base (NGA, 1991).[114] As states increasingly look toward regional solutions to managing hazardous waste, a manifest system becomes a more important planning tool.

Many states use manifest information for state-specific purposes, such as generating hazardous waste fees, meeting state reporting and management requirements, identifying potentially responsible parties for contaminated site cleanups, responding to public inquiries, and sharing information with other states. The 1991 NGA document provides states with a handbook to use in developing and maintaining an effective manifest tracking program. The guide is also a reference for planners wishing to find out how states process hazardous waste shipping information using the manifest.

Appendix V.1 EPA Hazardous Waste Codes

Code	Waste description
D001	Ignitable waste
D002	Corrosive waste
D003	Reactive waste
D004	Arsenic
D005	Barium
D006	Cadmium
D007	Chromium
D008	Lead
D009	Mercury
D010	Selenium
D011	Silver
D012	Endrin(1,2,3,4,10,10-hexachloro-1,7-epoxy-1,4,4a,5,6,7,8,8a-octahydro-1,4-endo, endo-5,8-dimeth-ano-naphthalene)
D013	Lindane (1,2,3,4,5,6-hexa-chlorocyclohexane, gamma isomer)
D014	Methoxychlor (1,1,1-trichloro-2,2-bis [p-methoxyphenyl] ethane)
D015	Toxaphene (C_{10} H_{10} Cl_8, Technical chlorinated camphene, 67-69 percent chlorine)
D016	2,4-D (2,4-Dichlorophenoxyacetic acid)
D017	2,4,5-TP Silvex (2,4,5-Trichlorophenoxypropionic acid)
D018	Benzene
D019	Carbon tetrachloride
D020	Chlordane
D021	Chlorobenzene
D022	Chloroform
D023	o-Cresol
D024	m-Cresol
D025	p-Cresol
D026	Cresol
D027	1,4-Dichlorobenzene
D028	1,2-Dichloroethane
D029	1,1-Dichloroethylene
D030	2,4-Dinitrotoluene
D031	Heptachlor (and its epoxide)
D032	Hexachlorobenzene
D033	Hexachlorobutadiene
D034	Hexachloroethane
D035	Methyl ethyl ketone
D036	Nitrobenzene
D037	Pentachlorophenol
D038	Pyridine
D039	Tetrachloroethylene
D040	Trichlorethylene
D041	2,4,5-Trichlorophenol
D042	2,4,6-Trichlorophenol
D043	Vinyl chloride

Appendix V.1 EPA Hazardous Waste Codes (Continued)

Code	Waste description
	HAZARDOUS WASTE FROM NONSPECIFIC SOURCES
F001	The following spent halogenated solvents used in degreasing: Tetrachloroethylene, trichlorethylene, methylene chloride, 1,1,1-trichloroethane, carbon tetrachloride and chlorinated fluorocarbons; all spent solvent mixtures/blends used in degreasing containing, before use, a total of ten percent or more (by volume) of one or more of the above halogenated solvents or those solvents listed in F002, F004, and F005; and still bottoms from the recovery of these spent solvents and spent solvent mixtures.
F002	The following spent halogenated solvents: Tetrachloroethylene, methylene chloride, trichloroethylene, 1,1,1-trichloroethane, chlorobenzene, 1,1,2-trichloro-1,2,2-trifluoroethane, ortho-dichlorobenzene, trichlorofluoromethane, and 1,1,2, trichloroethane; all spent solvent mixtures/blends containing, before use, a total of ten percent or more (by volume) of one or more of the above halogenated solvents or those solvents listed in F001, F004, and F005; and still bottoms from the recovery of these spent solvents and spent solvent mixtures.
F003	The following spent non-halogenated solvents: Xylene, acetone, ethyl acetate, ethyl benzene, ethyl ether, methyl isobutyl ketone, n-butyl alcohol, cyclohexanone, and methanol; all spent solvent mixtures/blends containing, before use, only the above spent nonhalogenated solvents; and all spent solvent mixtures/blends containing, before use, one or more of the above nonhalogenated solvents, and a total of ten percent or more (by volume) of one or more of those solvents listed in F001, F002, F004, and F005; and still bottoms from the recovery of these spent solvents and spent solvent mixtures.
F004	The following spent nonhalogenated solvents: cresols, cresylic acid, and nitrobenzene; and the still bottoms from the recovery of these solvents; all spent solvent mixtures/blends containing, before use, a total of ten percent or more (by volume) of one or more of the above nonhalogenated solvents or those solvents listed in F001, F002, and F005; and still bottoms from the recovery of these spent solvents and spent solvent mixtures.
F005	The following spent nonhalogenated solvents: toluene, methyl ethyl ketone, carbon disulfide, isobutanol, pyridine, benzene, 2-ethoxyethanol, and 2-nitropropane; all spent solvent mixtures/blends containing, before use, a total of ten percent or more (by volume) of one or more of the above nonhalogenated solvents or those solvents listed in F001, F002, or F004; and still bottoms from the recovery of these spent solvents and spent solvent mixtures.
F006	Wastewater treatment sludges from electroplating operations except from the following processes: (1) sulfuric acid anodizing of aluminum; (2) tin plating on carbon steel; (3) zinc plating (segregated basis) on carbon steel; (4) aluminum or zinc-aluminum plating on carbon steel; (5) cleaning/stripping associated with tin, zinc, and aluminum plating on carbon steel; and (6) chemical etching and milling of aluminum.
F007	Spent cyanide plating bath solutions from electroplating operations.
F008	Plating bath residues from the bottom of plating baths from electroplating operations in which cyanides are used in the process.
F009	Spent stripping and cleaning bath solutions from electroplating operations in which cyanides are used in the process.

Appendix V.1 EPA Hazardous Waste Codes (Continued)

Code	Waste description
F010	Quenching bath residues from oil baths from metal heat treating operations in which cyanides are used in the process.
F011	Spent cyanide solutions from slat bath pot cleaning from metal heat treating operations.
F012	Quenching wastewater treatment sludges from metal heat treating operations in which cyanides are used in the process.
F019	Wastewater treatment sludges from the chemical conversion coating of aluminum except from zirconium phosphating in aluminum can washing when such phosphating is an exclusive conversion coating process.
F020	Wastes (except wastewater and spent carbon from hydrogen chloride purification) from the production or manufacturing use (as a reactant, chemical intermediate, or component in a formulating process) of tri- or tetrachlorophenol or of intermediates used to produce their pesticide derivatives. (This listing does not include wastes from the production of hexachlorophene from highly purified 2,4,5-trichlorophenol.)
F021	Wastes (except wastewater and spent carbon from hydrogen chloride purification) from the production or manufacturing use (as a reactant, chemical intermediate, or component in a formulating process) of pentachlorophenol, or of intermediates used to produce derivatives.
F022	Wastes (except wastewater and spent carbon from hydrogen chloride purification) from the manufacturing use (as a reactant, chemical intermediate, or component in a formulating process) of tetra-, penta-, or hexachlorobenzenes under alkaline conditions.
F023	Wastes (except wastewater and spent carbon from hydrogen chloride purification) from the production of materials on equipment previously used for the production or manufacturing use (as a reactant, chemical intermediate, or component in a formulating process) of tri- and tetrachlorophenols. (This listing does not include wastes from equipment used only for the production or use of hexachlorophene from highly purified 2,4,5-trichlorophenol.)
F024	Process wastes including, but not limited to, distillation residues, heavy ends, tars, and reactor clean-out wastes, from the production of certain chlorinated aliphatic hydrocarbons by free radical catalyzed processes. These chlorinated aliphatic hydrocarbons are those having carbon chain lengths ranging from one to and including five, with varying amounts and positions of chlorine substitution. (This listing does not include wastewaters, wastewater treatment sludge, spent catalysts, and wastes listed in Sections 261.31. or 261.32)
F025	Condensed light ends, spent filters and filter aids, and spent desiccant wastes from the production of certain chlorinated aliphatic hydrocarbons, by free radical catalyzed processes. These chlorinated aliphatic hydrocarbons are those having carbon chain lengths ranging from one, to and including five, with varying amounts and positions of chlorine substitution.
F026	Wastes (except wastewater and spent carbon from hydrogen chloride purification) from the production of materials on equipment previously used for the manufacturing use (as a reactant, chemical intermediate, or component in a formulating process) of tetra-, penta-, or hexachlorobenzene under alkaline conditions.

Appendix V.1 EPA Hazardous Waste Codes (Continued)

Code	Waste description
F027	Discarded unused formulations containing tri-, tetra-, or pentachlorophenol or discarded unused formulations containing compounds derived from these chlorophenols. (This listing does not include formulations containing hexachlorophene synthesized from prepurified 2,4,5-trichlorophenol as the sole component.)
F028	Residues resulting from the incineration or thermal treatment of soil contaminated with EPA hazardous waste nos. F020, F021, F022, F023, F026, and F027.
F032	Wastewaters, process residuals, preservative drippage, and spent formulations from wood preserving processes generated at plants that currently use, or have previously used, chlorophenolic formulations [except potentially cross-contaminated wastes that have had the F032 waste code deleted in accordance with Section 261.35 (i.e., the newly promulgated equipment cleaning or replacement standards), and where the generator does not not resume or initiate use of chlorophenolic formulations]. (This listing does not include K001 bottom sediment sludge from the treatment of wastewater from wood preserving processes that use creosote and/or pentachlorophenol.)
F034	Wastewaters, process residuals, preservative drippage, and spent formulations from wood preserving processes generated at plants that use creosote formulations. This listing does not include K001 bottom sediment sludge from the treatment of wastewater from wood preserving processes that use creosote and/or pentachlorophenol.
F035	Wastewaters, process residuals, preservative drippage, and spent formulations from wood preserving processes generated at plants that use inorganic preservatives containing arsenic or chromium. This listing does not include K001 bottom sediment sludge from the treatment of wastewater from wood preserving processes that use creosote and/or pentachlorophenol.
F037	Petroleum refinery primary oil/water/solids separation sludge - Any sludge generated from the gravitational separation of oil/water/solids during the storage or treatment of process wastewaters and oily cooling wastewaters from petroleum refineries. Such sludges include, but are not limited to, those generated in oil/water/solids separators; tanks and impoundments; ditches and other conveyances; sumps; and storm water units receiving dry weather flow. Sludges generated in storm water units that do not receive dry weather flow, sludges generated in aggressive biological treatment units as defined in Section 261.31(b)(2)(including sludges generated in one or more additional units after wastewaters have been treated in aggressive biological treatment units), and K051 wastes are exempted from this listing.
F038	Petroleum refinery secondary (emulsified) oil/water/solids separation sludge - Any sludge and/or float generated from the physical and/or chemical separation of oil/water/solids in process wastewaters and oily cooling wastewaters from petroleum refineries. Such wastes include, but are not limited to, all sludges and floats generated in induced air flotation (IAF) units, tanks and impoundments, and all sludges generated in DAF units. Sludges generated in stormwater units that do not receive dry weather flow, sludges generated in aggressive biological treatment units as defined in Section

Appendix V.1 EPA Hazardous Waste Codes (Continued)

Code	Waste description
	261.31(b)(2) (including sludges generated in one or more additional units after wastewaters have been treated in aggressive biological treatment units), and F037, K048, and K051 wastes are exempted from this listing.
F039	Leachate resulting from the treatment, storage, or disposal of wastes classified by more than one waste code under Subpart D, or from a mixture of wastes classified under Subparts C and D of this part. (Leachate resulting from the management of one or more of the following EPA Hazardous Wastes and no other hazardous wastes retains its hazardous waste code(s): F020, F021, F022, F023, F026, F027, and/or F028.)
HAZARDOUS WASTE FROM SPECIFIC SOURCES	
K001	Bottom sediment sludge from the treatment of wastewaters from wood preserving processes that use creosote and/or pentachlorophenol.
K002	Wastewater treatment sludge from the production of chrome yellow and orange pigments.
K003	Wastewater treatment sludge from the production of molybdate orange pigments.
K004	Wastewater treatment sludge from the production of zinc yellow pigments.
K005	Wastewater treatment sludge from the production of chrome green pigments.
K006	Wastewater treatment sludge from the production of chrome oxide green pigments (anhydrous and hydrated).
K007	Wastewater treatment sludge from the production of iron blue pigments.
K008	Oven residue from the production of chrome oxide green pigments.
K009	Distillation bottoms from the production of acetaldehyde from ethylene.
K010	Distillation side cuts from the production of acetaldehyde from ethylene.
K011	Bottom stream from the wastewater stripper in the production of acrylonitrile.
K013	Bottom stream from the acetonitrile column in the production of acrylonitrile.
K014	Bottoms from the acetonitrile purification column in the production of acrylonitrile.
K015	Still bottoms from the distillation of benzyl chloride.
K016	Heavy ends or distillation residues from the production of carbon tetrachloride.
K017	Heavy ends (still bottoms) from the purification column in the production of epichlorohydrin.
K018	Heavy ends from the fractionation column in ethyl chloride production.
K019	Heavy ends from the distillation of ethylene dichloride in ethylene dichloride production.
K020	Heavy ends from the distillation of vinyl chloride in vinyl chloride monomer production.
K021	Aqueous spent antimony catalyst waste from fluoromethane production.
K022	Distillation bottom tars from the production of phenol/acetone from cumene.
K023	Distillation light ends from the production of phthalic anhydride from naphthalene.
K024	Distillation bottoms from the production of phthalic anhydride from naphthalene.

Appendix V.1 EPA Hazardous Waste Codes (Continued)

Code	Waste description
K025	Distillation bottoms from the production of nitrobenzene by the nitration of benzene.
K026	Stripping still tails from the production of methyl ethyl pyridines.
K027	Centrifuge and distillation residues from toluene diisocyanate production.
K028	Spent catalyst from the hydrochlorinator reactor in the production of 1,1,1-trichloroethane.
K029	Waste from the product steam stripper in the production of 1,1,1-trichloroethane.
K030	Column bottoms or heavy ends from the combined production of trichloroethylene and perchloroethylene.
K031	By-product salts generated in the production of MSMA and cacodylic acid.
K032	Wastewater treatment sludge from the production of chlordane.
K033	Wastewater and scrub water from the chlorination of cyclopentadiene in the production of chlordane.
K034	Filter solids from the filtration of hexachlorocyclopentadiene in the production of chlordane.
K035	Wastewater treatment sludges generated in the production of creosote.
K036	Still bottoms from toluene reclamation distillation in the production of disulfoton.
K037	Wastewater treatment sludges from the production of disulfoton.
K038	Wastewater from the washing and stripping of phorate production.
K039	Filter cake from the filtration of diethylphosphorodithioic acid in the production of phorate.
K040	Wastewater treatment sludge from the production of phorate.
K041	Wastewater treatment sludge from the production of toxaphene.
K042	Heavy ends or distillation residues from the distillation of tetrachlorobenzene in the production of 2,4,5-T.
K043	2,6-dichlorophenol waste from the production of 2,4-D.
K044	Wastewater treatment sludges from the manufacturing and processing of explosives.
K045	Spent carbon from the treatment of wastewater containing explosives.
K046	Wastewater treatment sludges from the manufacturing, formulation, and loading of lead-based initiating compounds.
K047	Pink/red water from TNT operations.
K048	Dissolved air flotation (DAF) float from the petroleum refining industry.
K049	Slop oil emulsion solids from the petroleum refining industry.
K050	Heat exchanger bundle cleaning sludge from the petroleum refining industry.
K051	API separator sludge from the petroleum refining industry.
K052	Tank bottoms (leaded) from the petroleum refining industry.
K060	Ammonia still lime sludge from coking operations.
K061	Emission control dust/sludge from the primary production of steel in electric furnaces.

Appendix V.1 EPA Hazardous Waste Codes (Continued)

Code	Waste description
K062	Spent pickle liquor from steel finishing operations of plants that produce iron or steel.
K064	Acid plant blowdown slurry/sludge resulting from the thickening of blowdown slurry from primary copper production.
K065	Surface impoundment solids contained in and dredged from surface impoundments at primary lead smelting facilities.
K066	Sludge from treatment of process wastewater and/or acid plant blowdown from primary zinc production.
K069	Emission control dust/sludge from secondary lead smelting.
K071	Brine purification muds from the mercury cell process in chlorine production, in which separately prepurified brine is not used.
K073	Chlorinated hydrocarbon waste from the purification step of the diaphragm cell process using graphite anodes in chlorine production.
K083	Distillation bottoms from aniline production.
K084	Wastewater treatment sludges generated during the production of veterinary pharmaceuticals from arsenic or organo-arsenic compounds.
K085	Distillation or fractionation column bottoms from the production of chlorobenzenes.
K086	Solvent washes and sludges, caustic washes and sludges, or water washes and sludges from cleaning tubs and equipment used in the formulation of ink from pigments, driers, soaps, and stabilizers containing chromium and lead.
K087	Decanter tank tar sludge from coking operations.
K088	Spent potliners from primary aluminum reduction.
K090	Emission control dust or sludge from ferrochromiumsilicon production.
K091	Emission control dust or sludge from ferrochromium production.
K093	Distillation light ends from the production of phthalic anhydride from ortho-xylene.
K094	Distillation bottoms from the production of phthalic anhydride from ortho-xylene.
K095	Distillation bottoms from the production of 1,1,1-trichloroethane.
K096	Heavy ends from the heavy ends column from the production of 1,1,1-trichloroethane.
K097	Vacuum stripper discharge from the chlordane chlorinator in the production of chlordane.
K098	Untreated process wastewater from the production of toxaphene.
K099	Untreated wastewater from the production of 2,4-D.
K100	Waste leaching solution from acid leaching of emission control dust/sludge from secondary lead smelting.
K101	Distillation tar residues from the distillation of aniline-based compounds in the production of veterinary pharmaceuticals from arsenic or organo-arsenic compounds.
K102	Residue from the use of activated carbon for decolorization in the production of veterinary pharmaceuticals from arsenic or organo-arsenic compounds.

Appendix V.1 EPA Hazardous Waste Codes (Continued)

Code	Waste description
K103	Process residues from aniline extraction from the production of aniline.
K104	Combined wastewaters generated from nitrobenzene/aniline production.
K105	Separated aqueous stream from the reactor product washing step in the production of chlorobenzenes.
K106	Wastewater treatment sludge from the mercury cell process in chlorine production.
K107	Column bottoms from product separation from the production of 1,1-dimethylhydrazine (UDMH) from carboxylic acid hydrazides.
K108	Condensed column overheads from product separation and condensed reactor vent gases from the production of 1,1-dimethylhydrazine from carboxylic acid hydrazides.
K109	Spent filter cartridges from product purification from the product of 1,1-dimethylhydrazine from carboxylic acid hydrazides.
K110	Condensed column overheads from intermediate separation from the production of 1,1-dimethylhydrazine from carboxylic acid hydrazides.
K111	Product washwaters from the production of dinitrotoluene via nitration of toluene.
K112	Reaction by-product water from the drying column in the production of toluenediamine via hydrogenation of dinitrotoluene.
K113	Condensed liquid light ends from purification of toluenediamine in production of toluenediamine via hydrogenation of dinitrotoluene.
K114	Vicinals from the purification of toluenediamine in production of toluenediamine via hydrogenation of dinitrotoluene.
K115	Heavy ends from purification of toluenediamine in the production of toluenediamine via hydrogenation of dinitrotoluene.
K116	Organic condensate from the solvent recovery column in the production of toluene diisocyanate via phosgenation of toluenediamine.
K117	Wastewater from the reactor vent gas scrubber in the production of ethylene dibromide via bromination of ethene.
K118	Spent adsorbent solids from purification of ethylene dibromide in the production of ethylene dibromide via bromination of ethene.
K123	Process wastewater (including supernates, filtrates, and washwaters) from the production of ethylenebisdithiocarbamic acid and its salts.
K124	Reactor vent scrubber water from the production of ethylenebisdithiocarbamic acid and its salts.
K125	Filtration, evaporation, and centrifugation solids from the production of ethylenebisdithiocarbamic acid and its salts.
K126	Baghouse dust and floor sweepings in milling and packaging operations from production or formulation of ethylenebisdithiocarbamic acid and its salts.
K131	Wastewater from the reactor and spent sulfuric acid from the acid dryer from the production of methyl bromide.

Appendix V.1 EPA Hazardous Waste Codes (Continued)

Code	Waste description
K132	Spent absorbent and wastewater separator solids from the production of methyl bromide.
K136	Still bottoms from the purification of ethylene dibromide in the production of ethylene dibromide via bromination of ethene.
K141	Process residues from the recovery of coal tar, including, but not limited to, tar collecting sump residues from the production of coke from coal or the recovery of coke by-products produced from coal. This listing does not include K087 (decanter tank sludge from coking operations).
K142	Tank storage residues from the production of coke from coal or from the recovery of coke by-products from coal.
K143	Process residues from the recovery of light oil, including, but not limited to, those generated in stills, decanters, and wash oil recovery units from the recovery of coke by-products produced from coal.
K144	Wastewater sump residues from light oil refining, including, but not limited to, intercepting or contamination sump sludges from the recovery of coke by-products produced from coal.
K145	Residues from naphthalene collection and recovery operations from the recovery of coke by-products produced from coal.
K147	Tar storage residues from coal tar refining.
K148	Residues from coal tar distillation, including, but not limited to, still bottoms.
K149	Distillation bottoms from the production of alpha (or methyl-) chlorinated tolunes, ring-chlorinated tolunes, benzol chlorides, and compounds with mixtures of these functional groups. [This waste does not include still bottoms from the distillation of benzol chloride]
K150	Organic residues excluding spent carbon adsorbent, from the spent chlorine gas and hydrochloric acid recovery processes associated with the production of alpha (or methyl-) chlorinated tolunes, benzol chlorides, and compounds with mixtures of these functional groups.
K151	Wastewater treatment sludges, excluding neutralization and biological sludges, generated during the treatment of wastewaters from the production of alpha (or methyl-) chlorinated tolunes, benzol chlorides, and compounds with mixtures of these functional groups.

DISCARDED COMMERCIAL CHEMICAL PRODUCTS, OFF-SPECIFICATION SPECIES, CONTAINER RESIDUALS, AND SPILL RESIDUES THEREOF–ACUTE HAZARDOUS WASTE

(AN ALPHABETIZED LISTING CAN BE FOUND AT 40 CFR 261.33.)

Code	Waste description
P001	2H-1-Benzopyran-2-one, 4-hydroxy-3-(3-oxo-1-phenylbutyl)-, & salts, when present at concentrations greater than 0.3%
P001	Warfarin, & salts, when present at concentrations greater than 0.3%
P002	1-Acetyl-2-thiourea
P002	Acetamide, N-(aminothioxomethyl)-
P003	2-Propenal
P003	Acrolein
P004	1,4,5,8-Dimethanonaphthalene, 1,2,3,4,10,10-hexa-chloro-1,4,4a,5,8,8a,-hexahydro-, (1alpha, 4alpha, 4abeta, 5alpha, 8alpha, 8abeta)-

Appendix V.1 EPA Hazardous Waste Codes (Continued)

Code	Waste description	Code	Waste description
P004	Aldrin	P020	Dinoseb
P005	2-Propen-1-ol	P020	Phenol, 2-(1-methylpropyl)-4,6-dinitro-
P005	Allyl alcohol	P021	Calcium cyanide
P006	Aluminum phosphide (R,T)	P021	Calcium cyanide $Ca(CN)_2$
P007	3(2H)-Isoxazolone, 5-(aminomethyl)-	P022	Carbon disulfide
P007	5-(Aminomethyl)-3-isoxazolol	P023	Acetaldehyde, chloro-
P008	4-Aminopyridine	P023	Chloroacetaldehyde
P008	4-Pyridinamine	P024	Benzenamine, 4-chloro-
P009	Ammonium picrate (R)	P024	p-Chloraniline
P009	Phenol, 2,4,6-trinitro-, ammonium salt (R)	P026	1-(o-Chlorophenyl)thiourea
P010	Arsenic acid H_3AsO_4	P026	Thiourea, (2-chlorophenyl)-
P011	Arsenic oxide As_2O_5	P027	3-Chloropropionitrile
P011	Arsenic pentoxide	P027	Propanenitrile, 3-chloro-
P012	Arsenic oxide As_2O_3	P028	Benzene, (chloromethyl)-
P012	Arsenic trioxide	P028	Benzyl chloride
P013	Barium cyanide	P029	Copper cyanide
P014	Benzenethiol	P029	Copper cyanide Cu(CN)
P014	Thiophenol	P030	Cyanides (soluble cyanide salts), not otherwise specified
P015	Beryllium	P031	Cyanogen
P016	Dichloromethyl ether	P031	Ethanedinitrile
P016	Methane, oxybis[chloro-	P033	Cyanogen chloride
P017	2-Propanone, 1-bromo-	P033	Cyanogen chloride (CN)Cl
P017	Bromoacetone	P034	2-Cyclohexyl-4,6-dinitrophenol
P018	Brucine	P034	Phenol, 2-cyclohexyl-4,6-dinitro-
P018	Strychnidin-10-one, 2,3-dimethoxy-		

Appendix V.1 EPA Hazardous Waste Codes (Continued)

Code	Waste description
P036	Arsonous dichloride, phenyl-
P036	Dichlorophenylarsine
P037	2,7:3,6-Dimethanonaphth[2,3-b]oxirene, 3,4,5,6,9,9-hexachloro-1a,2,2a,3,6,6a,7,7a-octahydro-, (1aalpha, 2beta, 2aalpha, 3beta, 6beta, 6aalpha, 7beta, 7aalpha)-
P037	Dieldrin
P038	Arsine, diethyl-
P038	Diethylarsine
P039	Disulfoton
P039	Phosphorodithioic acid, O,O-diethyl S-[2-(ethylthio)ethyl] ester
P040	O,O-Diethyl O-pyrazinyl phosphorothioate
P040	Phosphorothioic acid, O,O-diethyl O-pyrazinyl ester
P041	Diethyl-p-nitrophenyl phosphate
P041	Phosphoric acid, diethyl 4-nitrophenyl ester
P042	1,2-Benzenediol, 4-[1-hydroxy-2-(methylamino)ethyl]-, (R)-
P042	Epinephrine
P043	Diisopropylfluorophosphate (DFP)
P043	Phosphorofluoridic acid, bis(1-methylethyl) ester
P044	Dimethoate
P044	Phosphorodithioic acid, O,O-dimethyl S-[2-(methylamino)-2-oxoethyl] ester
P045	2-Butanone, 3,3-dimethyl-1-(methylthio)-, O-[methylamino)carbonyl] oxime
P045	Thiofanox
P046	alpha,alpha-Dimethylphenethylamine
P046	Benzeneethanamine, alpha, alpha-dimethyl-
P047	4,6-Dinitro-o-cresol, & salts
P047	Phenol, 2-methyl-4,6-dinitro-, & salts
P048	2,4-Dinitrophenol
P048	Phenol, 2,4-dinitro-
P049	Dithiobiuret
P049	Thioimidodicarbonic diamide $[(H_2N)C(S)]_2NH$
P050	6,9-Methano-2,4,3-benzodioxathiepin,6,7,8,9,10,10-hexachloro-1,5,5a,6,9,9a-hexahydro-,3-oxide
P050	Endosulfan
P051	2,7:3,6-Dimethanonaphth[2,3-b]oxirene, 3,4,5,6,9,9-hexachloro-1a,2,2a,3,6,6a,7,7a-octahydro-, (1aalpha, 2beta, 2abeta, 3alpha, 6alpha, 6abeta, 7beta, 7aalpha)- & metabolites
P051	Endrin
P051	Endrin, & metabolites
P054	Aziridine
P054	Ethyleneimine
P056	Fluorine
P057	Acetamide, 2-fluoro-
P057	Fluoroacetamide
P058	Acetic acid, fluoro-, sodium salt
P058	Fluoroacetic acid, sodium salt

Appendix V.1 EPA Hazardous Waste Codes (Continued)

Code	Waste description
P059	4,7-Methano-1H-indene, 1,4,5,6,7,8,8-heptachloro-3a,4,7,7a-tetrahydro-
P059	Heptachlor
P060	1,4,5,8-Dimethanonaphthalene, 1,2,3,4,10,10-hexa-chloro-1,4,4a,5,8,8a,-hexahydro-, (1alpha, 4alpha, 4abeta, 5beta, 8beta, 8abeta)-
P060	Isodrin
P062	Hexaethyl tetraphosphate
P062	Tetraphosphoric acid, hexaethyl ester
P063	Hydrocyanic acid
P063	Hydrogen cyanide
P064	Methane, isocyanato-
P064	Methyl isocyanate
P065	Fulminic acid, mercury(2+) salt (R,T)
P065	Mercury fulminate (R,T)
P066	Ethanimidothioic acid, N-[[(methylamino)carbonyl]oxy]-, methyl ester
P066	Methomyl
P067	1,2-Propylenimine
P067	Aziridine, 2-methyl-
P068	Hydrazine, methyl-
P068	Methyl hydrazine
P069	2-Methyllactonitrile
P069	Propanenitrile, 2-hydroxy-2-methyl-
P070	Aldicarb
P070	Propanal, 2-methyl-2-(methylthio)-, O-[(methylamino)carbonyl]oxime
P071	Methyl parathion
P071	Phosphorothioic acid, O,O,-dimethyl O-(4-nitrophenyl) ester
P072	alpha-Naphthylthiourea
P072	Thiourea, 1-naphthalenyl-
P073	Nickel carbonyl
P073	Nickel carbonyl $Ni(CO)_4$, (T-4)-
P074	Nickel cyanide
P074	Nickel cyanide $Ni(CN)_2$
P075	Nicotine, & salts
P075	Pyridine, 3-(1-methyl-2-pyrrolidinyl)-,(S)-, & salts
P076	Nitric oxide
P076	Nitrogen oxide NO
P077	Benzenamine, 4-nitro-
P077	p-Nitroaniline
P078	Nitrogen dioxide
P078	Nitrogen oxide NO_2
P081	1,2,3-Propanetriol, trinitrate (R)
P081	Nitroglycerine (R)
P082	Methanimine, N-methyl-N-nitroso-
P082	N-Nitrosodimethylamine
P084	N-Nitrosomethylvinylamine
P084	Vinylamine, N-methyl-N-nitroso-

Appendix V.1 EPA Hazardous Waste Codes (Continued)

Code	Waste description
P085	Diphosphoramide, octamethyl-
P085	Octamethylpyrophosphoramide
P087	Osmium oxide OsO_4, (T-4)-
P087	Osmium tetroxide
P088	7-Oxabicyclo[2.2.1]heptane-2,3-dicarboxylic acid
P088	Endothall
P089	Parathion
P089	Phosphorothioic acid, O,O-diethyl-O-(4-nitrophenyl) ester
P092	Mercury, (acetato-O)phenyl-
P092	Phenylmercury acetate
P093	Phenylthiourea
P093	Thiourea, phenyl-
P094	Phorate
P094	Phosphorodithioic acid, O,O-diethyl S-[(ethylthio)methyl] ester
P095	Carbonic dichloride
P095	Phosgene
P096	Hydrogen phosphide
P096	Phosphine
P097	Famphur
P097	Phosphorothioic acid O-[4-[(dimethylamino)sulfonyl]phenyl] O,O-dimethyl ester
P098	Potassium cyanide
P098	Potassium cyanide K(CN)
P099	Argentate (1-), bis(cyano-C)-, potassium
P099	Potassium silver cyanide
P101	Ethyl cyanide
P101	Propanenitrile
P102	2-Propyn-1-ol
P102	Propargyl alcohol
P103	Selenourea
P104	Silver cyanide
P104	Silver cyanide Ag(CN)
P105	Sodium azide
P106	Sodium cyanide
P106	Sodium cyanide Na(CN)
P107	Strontium sulfide SrS
P108	Strychnidin-10-one, & salts
P108	Strychnine, & salts
P109	Tetraethyldithiopyrophosphate
P109	Thiodiphosphoric acid, tetraethyl ester
P110	Plumbane, tetraethyl-
P110	Tetraethyl lead
P111	Diphosphoric acid, tetraethyl ester
P111	Tetraethyl pyrophosphate
P112	Methane, tetranitro- (R)
P112	Tetranitromethane (R)
P113	Thallic oxide
P113	Thallium oxide Tl_2O_3

Appendix V.1 EPA Hazardous Waste Codes (Continued)

Code	Waste description
P114	Selenious acid, dithallium (1+) salt
P114	Thallium(I) selenite
P115	Sulfuric acid, dithallium (1+) salt
P115	Thallium(I) sulfate
P116	Hydrazinecarbothioamide
P116	Thiosemicarbazide
P118	Methanethiol, trichloro-
P118	Trichloromethanethiol
P119	Ammonium vanadate
P119	Vanadic acid, ammonium salt
P120	Vanadium oxide V_2O_5
P120	Vanadium pentoxide
P121	Zinc cyanide
P121	Zinc cyanide $Zn(CN)_2$
P122	Zinc phosphide Zn_3P_2, when present at concentrations greater than 10% (R,T)
P123	Toxaphene

DISCARDED COMMERCIAL CHEMICAL PRODUCTS, OFF-SPECIFICATION SPECIES, CONTAINER RESIDUES, AND SPILL RESIDUES THEREOF–TOXIC WASTES

(AN ALPHABETIZED LISTING CAN BE FOUND AT 40 CFR 261.33.)

Code	Waste description
See F027	2,3,4,6-Tetrachlorophenol
	2,4,5-Tetrachlorophenol
	2,4,5-Trichlorophenol
	2,4,6-Trichlorophenol
	Acetic acid, (2,4,5-trichlorophenoxy)-
	Pentachlorophenol
	Phenol, 2,3,4,6-tetrachloro-
	Phenol, 2,4,5-trichloro-
	Phenol, 2,4,6-trichloro-
	Phenol, pentachloro-
	Propanoic acid, 2-(2,4,5-trichlorophenoxy)-
	Silvex (2,4,5-TP)
U001	Acetaldehyde (I)
U001	Ethanol (I)
U002	2-Propanone (I)
U002	Acetone (I)
U003	Acetonitrile (I,T)
U004	Acetophenone
U004	Ethanone, 1-phenyl-

Appendix V.1 EPA Hazardous Waste Codes (Continued)

Code	Waste description
U005	2-Acetylaminofluorene
U005	Acetamide, N-9H-fluoren-2-yl
U006	Acetyl chloride (C,R,T)
U007	2-Propenamide
U007	Acrylamide
U008	2-Propenoic acid (I)
U008	Acrylic acid (I)
U009	2-Propenenitrile
U009	Acrylonitrile
U010	Azirino [2',3':3,4]pyrrolo[1,2-a]indole-4,7-dione, 6-amino-8-[[(aminocarbonyl)oxy]methyl]-1,1a,2,8,8a,8b-hexahydro-8a-methoxy-5-methyl-, [1aS-(1aalpha, 8beta, 8aalpha, 8balpha)]-
U010	Mitomycin C
U011	1H-1,2,4-Triazol-3-amine
U011	Amitrole
U012	Aniline (I,T)
U012	Benzenamine (I,T)
U014	Auramine
U014	Benzenamine, 4,4'-carbonimidoylbis[N,N-dimethyl-
U015	Azaserine
U015	L-Serine, diazoacetate (ester)
U016	Benz[c]acridine
U017	Benzal chloride
U017	Benzene, (dichloromethyl)-
U018	Benz[a]anthracene
U019	Benzene (I,T)
U020	Benzenesulfonic acid chloride (C,R)
U020	Benzenesulfonyl chloride (C,R)
U021	[1,1'-Biphenyl]-4,4'-diamine
U021	Benzidine
U022	Benzo[a]pyrene
U023	Benzene, (trichloromethyl)-
U023	Benzotrichloride (C,R,T)
U024	Dichloromethoxy ethane
U024	Ethane, 1,1'-[methylenebis(oxy)]bis[2-chloro-
U025	Dichloroethyl ether
U025	Ethane, 1,1'-oxybis[2-chloro-
U026	Chlornaphazin
U026	Naphthalenamine, N,N'-bis(2-chloroethyl)-
U027	Dichloroisopropyl ether
U027	Propane, 2,2'-oxybis[2-chloro-
U028	1,2-Benzenedicarboxylic acid, bis(2-ethylhexyl) ester
U028	Diethylhexyl phthalate
U029	Methane, bromo-
U029	Methyl bromide
U030	4-Bromophenyl phenyl ether
U030	Benzene, 1-bromo-4-phenoxy-
U031	1-Butanol (I)

Appendix V.1 EPA Hazardous Waste Codes (Continued)

Code	Waste description
U031	n-Butyl alcohol (I)
U032	Calcium chromate
U032	Chromic acid H_2CrO_4, calcium salt
U033	Carbon oxyfluoride (R,T)
U033	Carbonic difluoride
U034	Acetaldehyde, trichloro-
U034	Chloral
U035	Benzenebutanoic acid, 4-[bis(2-chloroethyl)amino]-
U035	Chlorambucil
U036	4,7-Methano-1H-indene, 1,2,4,5,6,7,8,8-octachloro-2,3,3a,4,7,7a-hexahydro-
U036	Chlordane, alpha & gamma isomers
U037	Benzene, chloro-
U037	Chlorobenzene
U038	Benzeneacetic acid, 4-chloro-alpha-(4-chlorophenyl)-alpha-hydroxy-, ethyl ester
U038	Chlorobenzilate
U039	p-Chloro-m-cresol
U039	Phenol, 4-chloro-3-methyl-
U041	Epichlorohydrin
U041	Oxirane, (chloromethyl)-
U042	2-Chloroethyl vinyl ether
U042	Ethene, (2-chloroethoxy)-
U043	Ethene, chloro-
U043	Vinyl chloride
U044	Chloroform
U044	Methane, trichloro-
U045	Methane, chloro- (I,T)
U045	Methyl chloride (I,T)
U046	Chloromethyl methyl ether
U046	Methane, chloromethoxy-
U047	beta-Chloronaphthalene
U047	Naphthalene, 2-chloro-
U048	o-Chlorophenol
U048	Phenol, 2-chloro-
U049	4-Chloro-o-toluidine, hydrochloride
U049	Benzenamine, 4-chloro-2-methyl-, hydrochloride
U050	Chrysene
U051	Creosote
U052	Cresol (Cresylic acid)
U052	Phenol, methyl-
U053	2-Butenal
U053	Crotonaldehyde
U055	Benzene, (1-methylethyl)- (I)
U055	Cumene (I)
U056	Benzene, hexahydro- (I)
U056	Cyclohexane (I)
U057	Cyclohexanone (I)
U058	2H-1,3,2-Oxazaphosphorin-2-amine, N,N-bis(2-chloroethyl)tetrahydro-, 2-oxide

Appendix V.1 EPA Hazardous Waste Codes (Continued)

Code	Waste description
U058	Cyclophosphamide
U059	5,12-Naphthacenedione, 8-acetyl-10-[(3-amino-2,3,6-trideoxy)-alpha-L-lyxo-hexopyranosyl)oxy]-7,8,9,10-tetrahydro-6,8,11-trihydroxy-1-methoxy-, (8S-cis)-
U059	Daunomycin
U060	Benzene, 1,1'-(2,2-dichloroethylidene)bis[4-chloro-
U060	DDD
U061	Benzene, 1,1'-(2,2,2-trichloroethylidene)bis[4-chloro-
U061	DDT
U062	Carbamothioic acid, bis(1-methylethyl)-, S-(2,3-dichloro-2-propenyl) ester
U062	Diallate
U063	Dibenz[a,h]anthracene
U064	Benzo[rst]pentaphene
U064	Dibenzo[a,i]pyrene
U066	1,2-Dibromo-3-chloropropane
U066	Propane, 1,2-dibromo-3-chloro-
U067	Ethane, 1,2-dibromo-
U067	Ethylene dibromide
U068	Methane, dibromo-
U068	Methylene bromide
U069	1,2-Benzenedicarboxylic acid, dibutyl ester
U069	Dibutyl phthalate
U070	Benzene, 1,2-dichloro-
U070	o-Dichlorobenzene
U071	Benzene, 1,3-dichloro-
U071	m-Dichlorobenzene
U072	Benzene, 1,4-dichloro-
U072	p-Dichlorobenzene
U073	[1,1'-Biphenyl]-4,4'-diamine, 3,3'-dichloro-
U073	3,3'-Dichlorobenzidine
U074	1,4-Dichloro-2-butene (I,T)
U074	2-Butene, 1,4-dichloro- (I,T)
U075	Dichlorodifluoromethane
U075	Methane, dichlorodifluoro-
U076	Ethane, 1,1-dichloro-
U076	Ethylidene dichloride
U077	Ethane, 1,2-dichloro-
U077	Ethylene dichloride
U078	1,1-Dichloroethylene
U078	Ethene, 1,1-dichloro-
U079	1,2-Dichloroethylene
U079	Ethene, 1,2-dichloro-,(E)-
U080	Methane, dichloro-
U080	Methylene chloride
U081	2,4-Dichlorophenol
U081	Phenol, 2,4-dichloro-
U082	2,6-Dichlorophenol
U082	Phenol, 2,6-dichloro-

Appendix V.1 EPA Hazardous Waste Codes (Continued)

Code	Waste description
U083	Propane, 1,2-dichloro-
U083	Propylene dichloride
U084	1,3-Dichloropropene
U084	1-Propene, 1,3-dichloro-
U085	1,2:3,4-Diepoxybutane (I,T)
U085	2,2'-Bioxirane
U086	Hydrazine, 1,2-diethyl-
U086	N,N'-Diethylhydrazine
U087	O,O-Diethyl S-methyl dithiophosphate
U087	Phosphorodithioic acid, O,O-diethyl S-methyl ester
U088	1,2-Benzenedicarboxylic acid, diethyl ester
U088	Diethyl phthalate
U089	Diethylstilbesterol
U089	Phenol, 4,4'-(1,2-diethyl-1,2-ethenediyl)bis, (E)-
U090	1,3-Benzodioxole, 5-propyl-
U090	Dihydrosafrole
U091	[1,1'-Biphenyl]-4,4'-diamine, 3,3'-dimethoxy-
U091	3,3'-Dimethoxybenzidine
U092	Dimethylamine (I)
U092	Methanamine, N-methyl- (I)
U093	Benzenamine, N,N-dimethyl-4-(phenylazo)-
U093	p-Dimethylaminoazobenzene
U094	7,12-Dimethylbenz[a]anthracene
U094	Benz[a]anthracene, 7,12-dimethyl-
U095	[1,1'-Biphenyl]-4,4'-diamine, 3,3'-dimethyl-
U095	3,3'-Dimethylbenzidine
U096	alpha,alpha-Dimethylbenzylhydroperoxide (R)
U096	Hydroperoxide, 1-methyl-1-phenylethyl- (R)
U097	Carbamic chloride, dimethyl-
U097	Dimethylcarbamoyl chloride
U098	1,1-Dimethylhydrazine
U098	Hydrazine, 1,1-dimethyl-
U099	1,2-Dimethylhydrazine
U099	Hydrazine, 1,2-diphenyl-
U101	2,4-Dimethylphenol
U101	Phenol, 2,4-dimethyl-
U102	1,2-Benzenedicarboxylic acid, dimethyl ester
U102	Dimethyl phthalate
U103	Dimethyl sulfate
U103	Sulfuric acid, dimethyl ester
U105	2,4-Dinitrotoluene
U105	Benzene, 1-methyl-2,4-dinitro-
U106	2,6-Dinitrotoluene
U106	Benzene, 2-methyl-1,3-dinitro-
U107	1,2-Benzenedicarboxylic acid, dioctyl ester
U107	Di-n-octyl phthalate

Appendix V.1 EPA Hazardous Waste Codes (Continued)

Code	Waste description
U108	1,4-Diethyleneoxide
U108	1,4-Dioxane
U109	1,2-Diphenylhydrazine
U109	Hydrazine, 1,2-diphenyl-
U110	1-Propanimine, N-propyl-(I)
U110	Dipropylamine (I)
U111	1-Propanamine, N-nitroso-N-propyl-
U111	Di-n-propylnitrosamine
U112	Acetic acid, ethyl ester (I)
U112	Ethyl acetate (I)
U113	2-Propenoic acid, ethyl ester (I)
U113	Ethyl acrylate (I)
U114	Carbamodithioic acid, 1,2-ethanediylbis-, salts & esters
U114	Ethylenebisdithiocarbamic acid, salts & esters
U115	Ethylene oxide (I,T)
U115	Oxirane (I,T)
U116	2-Imidazolidinethione
U116	Ethylenethiourea
U117	Ethane, 1,1'-oxybis-(I)
U117	Ethyl ether (I)
U118	2-Propenoic acid, 2-methyl-, ethyl ester
U118	Ethyl methacrylate
U119	Ethyl methanesulfonate
U119	Methanesulfonic acid, ethyl ester
U120	Fluoranthene
U121	Methane, trichlorofluoro-
U121	Trichloromonofluoromethane
U122	Formaldehyde
U123	Formic acid (C,T)
U124	Furan (I)
U124	Furfuran (I)
U125	2-Furancarboxaldehyde (I)
U125	Furfural (I)
U126	Glycidylaldehyde
U126	Oxiranecarboxyaldehyde
U127	Benzene, hexachloro-
U127	Hexachlorobenzene
U128	1,3-Butadiene, 1,1,2,3,4,4-hexachloro-
U128	Hexachlorobutadiene
U129	Cyclohexane, 1,2,3,4,5,6-hexachloro-, (1alpha, 2alpha, 3beta, 4alpha, 5alpha, 6beta)-
U129	Lindane
U130	1,3-Cyclopentadiene, 1,2,3,4,5,5-hexachloro-
U130	Hexachlorocyclopentadiene
U131	Ethane, hexachloro-
U131	Hexachloroethane
U132	Hexachlorophene
U132	Phenol, 2,2 -methylenebis[3,4,6-trichloro-

Appendix V.1 EPA Hazardous Waste Codes (Continued)

Code	Waste description
U133	Hydrazine (R,T)
U134	Hydrofluoric acid (C,T)
U134	Hydrogen fluoride (C,T)
U135	Hydrogen sulfide
U135	Hydrogen sulfide H_2S
U136	Arsenic acid, dimethyl-
U136	Cacodylic acid
U137	Indeno[1,2,3-cd]pyrene
U138	Methane, iodo-
U138	Methyl iodide
U140	1-Propanol, 2-methyl- (I,T)
U140	Isobutyl alcohol (I,T)
U141	1,3-Benzodioxole, 5-(1-propenyl)-
U141	Isosafrole
U142	1,3,4-Metheno-2H-cyclobuta[cd]pentalen-2-one, 1,1a,3,3a,4,5,5,5a,5b,6-decachlorooctahydro-
U142	Kepone
U143	2-Butenoic acid, 2-methyl-, 7-[[2,3-dihydroxy-2-(1-methoxyethyl)-3-methyl-1-oxobutoxy]methyl]-2,3,5,7a-tetrahydro-1H-pyrrolizin-1-yl ester, [1S-[1alpha(Z), 7(2S*,3R*), 7aalpha]]-
U143	Lasiocarpine
U144	Acetic acid, lead(2+) salt
U144	Lead acetate
U145	Lead phosphate
U145	Phosphoric acid, lead(2+) salt (2:3)
U146	Lead subacetate
U146	Lead, bis(acetato-O)tetrahydroxytri-
U147	2,5-Furandione
U147	Maleic anhydride
U148	3,6-Pyridazinedione, 1,2-dihydro-
U148	Maleic hydrazide
U149	Malononitrile
U149	Propanedinitrile
U150	L-Phenylalanine, 4-[bis(2-chloroethyl)amino]-
U150	Melphalan
U151	Mercury
U152	2-Propenenitrile, 2-methyl- (I,T)
U152	Methacrylonitrile (I,T)
U153	Methanethiol (I,T)
U153	Thiomethanol (I,T)
U154	Methanol (I)
U154	Methyl alcohol (I)
U155	1,2-Ethanediamine, N,N-dimethyl-N'-2-pyridinyl-N'-(2-thienylmethyl)-
U155	Methapyrilene
U156	Carbonochloridic acid, methyl ester, (I,T)
U156	Methyl chlorocarbonate (I,T)
U157	3-Methylcholanthrene
U157	Benz[j]aceanthrylene, 1,2-dihydro-3-methyl-

Appendix V.1 EPA Hazardous Waste Codes (Continued)

Code	Waste description
U158	4,4'-Methylenebis(2-chloroaniline)
U158	Benzenamine, 4,4'-methylenebis[2-chloro-
U159	2-Butanone (I,T)
U159	Methyl ethyl ketone (MEK) (I,T)
U160	2-Butanone, peroxide (R,T)
U160	Methyl ethyl ketone peroxide (R,T)
U161	4-Methyl-2-pentanone (I)
U161	Methyl isobutyl ketone (I)
U161	Pentanol, 4-methyl-
U162	2-Propenoic acid, 2-methyl-, methyl ester (I,T)
U162	Methyl methacrylate (I,T)
U163	Guanidine, N-methyl-N'-nitro-N-nitroso-
U163	MNNG
U164	4(1H)-Pyrimidinone, 2,3-dihydro-6-methyl-2-thioxo-
U164	Methylthiouracil
U165	Naphthalene
U166	1,4-Naphthalenedione
U166	1,4-Naphthoquinone
U167	1-Naphthalenamine
U167	alpha-Naphthylamine
U168	2-Naphthalenamine
U168	beta-Naphthylamine
U169	Benzene, nitro-
U169	Nitrobenzene (I,T)
U170	p-Nitrophenol (I,T)
U170	Phenol, 4-nitro-
U171	2-Nitropropane (I,T)
U171	Propane, 2-nitro- (I,T)
U172	1-Butanamine, N-butyl-N-nitroso-
U172	N-Nitrosodi-n-butylamine
U173	Ethanol, 2,2'-(nitrosoimino)bis-
U173	N-Nitrosodiethanolamine
U174	Ethanamine, N-ethyl-N-nitroso-
U174	N-Nitrosodiethylamine
U176	N-Nitroso-N-ethylurea
U176	Urea, N-ethyl-N-nitroso-
U177	N-Nitroso-N-methylurea
U177	Urea, N-methyl-N-nitroso-
U178	Carbamic acid, methylnitroso-, ethyl ester
U178	N-Nitroso-N-methylurethane
U179	N-Nitrosopiperidine
U179	Piperidine, 1-nitroso-
U180	N-Nitrosopyrrolidine
U180	Pyrrolidine, 1-nitroso-
U181	5-Nitro-o-toluidine
U181	Benzenamine, 2-methyl-5-nitro
U182	1,3,5-Trioxane, 2,4,6-trimethyl-
U182	Paraldehyde
U183	Benzene, pentachloro-

Appendix V.1 EPA Hazardous Waste Codes (Continued)

Code	Waste description
U183	Pentachlorobenzene
U184	Ethane, pentachloro-
U184	Pentachloroethane
U185	Benzene, pentachloronitro-
U185	Pentachloronitrobenzene (PCNB)
U186	1,3-Pentadiene (I)
U186	1-Methylbutadiene (I)
U187	Acetamide, N-(4-ethoxyphenyl)-
U187	Phenacetin
U188	Phenol
U189	Phosphorus sulfide (R)
U189	Sulfur phosphide (R)
U190	1,3-Isobenzofurandione
U190	Phthalic anhydride
U191	2-Picoline
U191	Pyridine, 2-methyl-
U192	Benzamide, 3,5-dichloro-N-(1,1-dimethyl-2-propynyl)-
U192	Pronamide
U193	1,2-Oxathiolane, 2,2-dioxide
U193	1,3-Propane sultone
U194	1-Propanamine (I,T)
U194	n-Propylamine (I,T)
U196	Pyridine
U197	2,5-Cyclohexadiene-1,4-dione
U197	p-Benzoquinone
U200	Reserpine
U200	Yohimban-16-carboxylic acid, 11,17-dimethoxy-18-[(3,4,5-trimethoxybenzoyl)oxy]-, methyl ester, (3beta, 16beta, 17alpha, 18beta, 20alpha)-
U201	1,3-Benzenediol
U201	Resorcinol
U202	1,2-Benzisothiazol-3(2H)-one, 1,1-dioxide, & salts
U202	Saccharin, & salts
U203	1,3-Benzodioxole, 5-(2-propenyl)-
U203	Safrole
U204	Selenious acid
U204	Selenium dioxide
U205	Selenium sulfide
U205	Selenium sulfide SeS_2 (R,T)
U206	D-Glucose, 2-deoxy-2-[[(methylnitrosoamino)-carbonyl]amino]-
U206	Glucopyranose, 2-deoxy-2-(3-methyl-3-nitrosoureido)-,D-
U206	Streptozotocin
U207	1,2,4,5-Tetrachlorobenzene
U207	Benzene, 1,2,4,5-tetrachloro-
U208	1,1,1,2-Tetrachloroethane
U208	Ethane, 1,1,1,2-tetrachloro-
U209	1,1,2,2-Tetrachloroethane
U209	Ethane, 1,1,2,2-tetrachloro-

Appendix V.1 EPA Hazardous Waste Codes (Continued)

Code	Waste description	Code	Waste description
U210	Ethene, tetrachloro-	U225	Bromoform
U210	Tetrachloroethylene	U225	Methane, tribromo-
U211	Carbon tetrachloride	U226	Ethane, 1,1,1-trichloro-
U211	Methane, tetrachloro-	U226	Methyl chloroform
U213	Furan, tetrahydro-(I)	U227	1,1,2-Trichloroethane
U213	Tetrahydrofuran (I)	U227	Ethane, 1,1,2-trichloro-
U214	Acetic acid, thallium(1+) salt	U228	Ethene, trichloro-
U214	Thallium(I) acetate	U228	Trichloroethylene
U215	Carbonic acid, dithallium(1+) salt	U234	1,3,5-Trinitrobenzene (R,T)
U215	Thallium(I) carbonate	U234	Benzene, 1,3,5-trinitro-
U216	Thallium chloride Tlcl	U235	1-Propanol, 2,3-dibromo-, phosphate (3:1)
U216	Thallium(I) chloride	U235	Tris(2,3,-dibromopropyl) phosphate
U217	Nitric acid, thallium(1+) salt	U236	2,7-Naphthalenedisulfonic acid,3,3'-[(3,3'-dimethyl[1,1'-biphenyl]-4,4'-diyl)bis(azo)bis[5-amino-4-hydroxy]-, tetrasodium salt
U217	Thallium(I) nitrate	U236	Trypan blue
U218	Ethanethioamide	U237	2,4-(1H,3H)-Pyrimidinedione, 5-[bis(2-chloroethyl)amino]-
U218	Thioacetamide	U237	Uracil mustard
U219	Thiourea	U238	Carbamic acid, ethyl ester
U220	Benzene, methyl-	U238	Ethyl carbamate (urethane)
U220	Toluene	U239	Benzene, dimethyl- (I,T)
U221	Benzenediamine, ar-methyl-	U239	Xylene (I)
U221	Toluenediamine	U240	2,4-D, salts & esters
U222	Benzenamine, 2-methyl-, hydrochloride	U240	Acetic acid, (2,4-dichlorophenoxy)-, salts & esters
U222	o-Toluidine hydrochloride		
U223	Benzene, 1,3-diisocyanatomethyl- (R,T)		
U223	Toluene diisocyanate (R,T)		

Appendix V.1 EPA Hazardous Waste Codes (Continued)

Code	Waste description	Code	Waste description
U243	1-Propene, 1,1,2,3,3,3-hexachloro-		
U243	Hexachloropropene		
U244	Thioperoxydicarbonic diamide $[(H_2N)C(S)]_2S_2$, tetramethyl-		
U244	Thiram		
U246	Cyanogen bromide (CN)Br		
U247	Benzene, 1,1'-(2,2,2-trichloroethylidene)bis[4-methoxy-		
U247	Methoxychlor		
U248	2H-1-Benzopyran-2-one, 4-hydroxy-3-(3-oxo-1-phenyl-butyl)-, & salts, when present at concentrations of 0.3% or less		
U248	Warfarin, & salts, when present at concentrations of 0.3% or less		
U249	Zinc phosphide Zn_3P_2, when present at concentrations of 10% or less		
U328	Benzenamine, 2-methyl-		
U328	o-Toluidine		
U353	Benzenamine, 4-methyl-		
U353	p-Toluidine		
U359	Ethanol, 2-ethoxy-		
U359	Ethylene glycol monoethyl ether		

Appendix V.2 Source Codes

Code	Waste source
CLEANING AND DEGREASING	
A01	Stripping
A02	Acid cleaning
A03	Caustic (Alkali) cleaning
A04	Flush rinsing
A05	Dip rinsing
A06	Spray rinsing
A07	Vapor degreasing
A08	Physical scraping and removal
A09	Clean out process equipment
A19	Other cleaning and degreasing
SURFACE PREPARATION AND FINISHING	
A21	Painting
A22	Electroplating
A23	Electroless plating
A24	Phosphating
A25	Heat treating
A26	Pickling
A27	Etching
A29	Other surface coating/preparation (Specify in Comments)
PROCESSES OTHER THAN SURFACE PREPARATION	
A31	Product rinsing
A32	Product filtering
A33	Product distillation
A34	Product solvent extraction
A35	By-product processing
A36	Spent catalyst removal
A37	Spent process liquids removal
A38	Tank sludge removal
A39	Slag removal
A40	Metal forming
A41	Plastics forming
A49	Processes other than surface preparation (Specify in Comments)
PRODUCTION OR SERVICE DERIVED ONE-TIME AND INTERMITTENT PROCESSES	
A51	Leak collection
A53	Cleanup of spill residues
A54	Oil changes
A55	Filter/Battery replacement
A56	Discontinue use of process equipment
A57	Discarding off-spec material
A58	Discarding out-of-date products or chemicals
A59	Other production-derived one-time and intermittent processes
A60	Sludge removal
REMEDIATION DERIVED WASTE	
A61	Superfund Remedial Action
A62	Superfund Emergency Response
A63	RCRA Corrective Action at solid waste management unit
A64	RCRA closure of hazardous waste management unit
A65	Underground storage tank cleanup
A69	Other remediation
POLLUTION CONTROL OR WASTE TREATMENT PROCESSES	
A71	Filtering/screening
A72	Metals recovery
A73	Solvents recovery
A74	Incineration/Thermal treatment
A75	Wastewater treatment
A76	Sludge dewatering
A77	Stabilization
A78	Air pollution control devices
A79	Leachate collection
A89	Other pollution control or waste treatment
OTHER PROCESSES	
A91	Clothing and personal protective equipment
A92	Routine cleanup wastes (e.g., floor sweepings)
A93	Closure of management unit(s) or equipment other than by remediation specified in codes A61 - A69
A94	Laboratory wastes
A99	Other

Appendix V.3 Form Codes

LAB PACKS

LAB PACKS - Lab packs of mixed wastes, chemicals, lab wastes

Code	Waste description
B001	Lab packs of old chemicals only
B002	Lab packs of debris only
B003	Mixed lab packs
B004	Lab packs containing acute hazardous wastes
B009	Other lab packs (Specify in Comments)

LIQUIDS

INORGANIC LIQUIDS - Waste that is primarily inorganic and highly fluid (e.g., aqueous), with low suspended inorganic solids and low organic content

Code	Waste description
B101	Aqueous waste with low solvents
B102	Aqueous waste with low other toxic organics
B103	Spent acid with metals
B104	Spent acid without metals
B105	Acidic aqueous waste
B106	Caustic solution with metals but no cyanides
B107	Caustic solution with metals and cyanides
B108	Caustic solution with cyanides but no metals
B109	Spent caustic
B110	Caustic aqueous waste
B111	Aqueous waste with reactive sulfides
B112	Aqueous waste with other reactives (e.g., explosives)
B113	Other aqueous waste with high dissolved solids
B114	Other aqueous waste with low dissolved solids
B115	Scrubber water
B116	Leachate
B117	Waste liquid mercury
B119	Other inorganic liquids (Specify in Comments)

ORGANIC LIQUIDS - Waste that is primarily organic and is highly fluid, with low inorganic solids content and low-to-moderate water content

Code	Waste description
B201	Concentrated solvent-water solution
B202	Halogenated (e.g., chlorinated) solvent
B203	Nonhalogenated solvent
B204	Halogenated/nonhalogenated solvent mixture
B205	Oil-water emulsion or mixture
B206	Waste oil
B207	Concentrated aqueous solution of other organics
B208	Concentrated phenolics
B209	Organic paint, ink, lacquer, or varnish
B210	Adhesives or epoxies
B211	Paint thinner or petroleum distillates
B212	Reactive or polymerizable organic liquid
B219	Other organic liquids (Specify in Comments)

SOLIDS

INORGANIC SOLIDS - Waste that is primarily inorganic and solid, with low organic content and low-to-moderate water content; not pumpable

Code	Waste description
B301	Soil contaminated with organics
B302	Soil contaminated with inorganics only
B303	Ash, slag, or other residue from incineration of wastes
B304	Other "dry" ash, slag, or thermal residue
B305	"Dry" lime or metal hydroxide solids chemically "fixed"
B306	"Dry" lime or metal hydroxide solids not "fixed"
B307	Metal scale, filings, or scrap
B308	Empty or crushed metal drums or containers
B309	Batteries or battery parts, casings, cores
B310	Spent solid filters or adsorbents
B311	Asbestos solids and debris
B312	Metal-cyanide salts/chemicals
B313	Reactive cyanide salts/chemicals
B314	Reactive sulfide salts/chemicals
B315	Other reactive salts/chemicals
B316	Other metal salts/chemicals
B319	Other waste inorganic solids (Specify in Comments)

ORGANIC SOLIDS - Waste that is primarily organic and solid, with low-to-moderate inorganic content and water content; not pumpable

Code	Waste description
B401	Halogenated pesticide solid
B402	Nonhalogenated pesticide solid
B403	Solid resins or polymerized organics
B404	Spent carbon
B405	Reactive organic solid
B406	Empty fiber or plastic containers

Appendix V.3 Form Codes (Continued)

Code	Waste description
B407	Other halogenated organic solids (Specify in Comments)
B409	Other nonhalogenated organic solids (Specify in Comments)

SLUDGES

INORGANIC SLUDGES - Waste that is primarily inorganic, with moderate-to-high water content and low organic content, and pumpable

Code	Waste description
B501	Lime sludge without metals
B502	Lime sludge with metals/metal hydroxide sludge
B503	Wastewater treatment sludge with toxic organics
B504	Other wastewater treatment sludge
B505	Untreated plating sludge without cyanides
B506	Untreated plating sludge with cyanides
B507	Other sludge with cyanides
B508	Sludge with reactive sulfides
B509	Sludge with other reactives
B510	Degreasing sludge with metal scale or filings
B511	Air pollution control device sludge (e.g., fly ash, wet scrubber sludge)
B512	Sediment or lagoon dragout contaminated with organics
B513	Sediment or lagoon dragout contaminated with inorganics only
B514	Drilling mud
B515	Asbestos slurry or sludge
B516	Chloride or other brine sludge
B519	Other inorganic sludges (Specify in Comments)

ORGANIC SLUDGES - Waste that is primarily organic with low-to-moderate inorganic solids content and water content, and pumpable

Code	Waste description
B601	Still bottoms of halogenated (e.g., chlorinated) solvents or other organic liquids
B602	Still bottoms of nonhalogenated solvents or other organic liquids
B603	Oily sludge
B604	Organic paint or ink sludge
B605	Reactive or polymerizable organics
B606	Resins, tars, or tarry sludge
B607	Biological treatment sludge
B608	Sewage or other untreated biological sludge
B609	Other organic sludges (Specify in Comments)

GASES

INORGANIC GASES - Waste that is primarily inorganic with a low organic content and is a gas at atmospheric pressure

Code	Waste description
B701	Inorganic gases

ORGANIC GASES - Waste that is primarily organic with low-to-moderate inorganic content and is a gas at atmospheric pressure

Code	Waste description
B801	Organic gases

Appendix V.4 System Type Codes

Code	System type
METALS RECOVERY (FOR REUSE)	
M011	High temperature metals recovery
M012	Retorting
M013	Secondary smelting
M014	Other metals recovery for reuse: e.g., ion exchange, reverse osmosis, acid leaching, etc. (Specify in Comments)
M019	Metals recovery - type unknown
SOLVENTS RECOVERY	
M021	Fractionation/distillation
M022	Thin film evaporation
M023	Solvent extraction
M024	Other solvent recovery (Specify in Comments)
M029	Solvents recovery - type unknown
OTHER RECOVERY	
M031	Acid regeneration
M032	Other recovery: e.g., waste oil recovery, nonsolvent organics recovery, etc. (Specify in Comments)
M039	Other recovery - type unknown
INCINERATION	
M041	Incineration - liquids
M042	Incineration - sludges
M043	Incineration - solids
M044	Incineration - gases
M049	Incineration - type unknown
ENERGY RECOVERY (REUSE AS FUEL)	
M051	Energy recovery - liquids
M052	Energy recovery - sludges
M053	Energy recovery - solids
M059	Energy recovery - type unknown
FUEL BLENDING	
M061	Fuel blending
AQUEOUS INORGANIC TREATMENT	
M071	Chrome reduction followed by chemical precipitation
M072	Cyanide destruction followed by chemical precipitation
M073	Cyanide destruction only
M074	Chemical oxidation followed by chemical precipitation
M075	Chemical oxidation only
M076	Wet air oxidation
M077	Chemical precipitation
M078	Other aqueous inorganic treatment: e.g., ion exchange, reverse osmosis, etc. (Specify in Comments)
M079	Aqueous inorganic treatment - type unknown
AQUEOUS ORGANIC TREATMENT	
M081	Biological treatment
M082	Carbon adsorption
M083	Air/steam stripping
M084	Wet air oxidation
M085	Other aqueous organic treatment (Specify in Comments)
M089	Aqueous organic treatment - type unknown
AQUEOUS ORGANIC AND INORGANIC TREATMENT	
M091	Chemical precipitation in combination with biological treatment
M092	Chemical precipitation in combination with carbon adsorption
M093	Wet air oxidation
M094	Other organic/inorganic treatment (Specify in Comments)
M099	Aqueous organic and inorganic treatment - type unknown
SLUDGE TREATMENT	
M101	Sludge dewatering
M102	Addition of excess lime
M103	Absorption/adsorption
M104	Solvent extraction
M109	Sludge treatment - type unknown

Appendix V.4 System Type Codes (Continued)

Code	System type	Code	System type
STABILIZATION			
M111	Stabilization/Chemical fixation using cementitious and/or pozzolanic materials		
M112	Other stabilization (Specify in Comments)		
M119	Stabilization - type unknown		
OTHER TREATMENT			
M121	Neutralization only		
M122	Evaporation only		
M123	Settling/clarification only		
M124	Phase separation (e.g., emulsion breaking, filtration) only		
M125	Other treatment (Specify in Comments)		
M129	Other treatment - type unknown		
DISPOSAL			
M131	Land treatment/application/farming		
M132	Landfill		
M133	Surface impoundment (to be closed as a landfill)		
M134	Deepwell/underground injection		
M135	Direct discharge to sewer/POTW (no prior treatment)		
M136	Direct discharge to surface water under NPDES (no prior treatment)		
M137	Other disposal (Specify in Comments)		
TRANSFER FACILITY STORAGE			
M141	Transfer facility storage, waste was shipped off site with no on-site TDR activity		

Appendix V.5 Activity Codes

Code	Waste minimization activity
	RECYCLING ACTIVITY
W01	On-site beneficial use/reuse began during 1993
W02	Off-site beneficial use/reuse began during 1993
	SOURCE REDUCTION ACTIVITY
	GOOD OPERATING PRACTICES
W11	Began to segregate types of hazardous waste to make them more amenable to recycling
W12	Began to segregate (stopped combining) hazardous waste from non-hazardous waste (Note: for purposes of hazardous waste reporting, reduces volume of hazardous waste, but does not reduce total waste volume)
W13	Improved maintenance scheduling, recordkeeping, or procedures
W14	Changed production schedule to minimize equipment and feedstock changeovers
W19	Other changes in operating practices (Specify in Comments)
	INVENTORY CONTROL
W21	Instituted procedures to ensure that materials do not stay in inventory beyond shelf-life
W22	Began to test outdated material--continue to use if still effective
W23	Eliminated shelf-life requirements for stable materials
W24	Instituted better labelling procedures
W25	Instituted clearinghouse to exchange materials that would otherwise be discarded
W29	Other (Specify in Comments)
	SPILL AND LEAK PREVENTION
W31	Improved storage or stacking procedures
W32	Improved procedures for loading, unloading, and t ransfer operations
W33	Installed overflow alarms or automatic shut-off valves
W34	Installed secondary containment
W35	Installed vapor recovery systems
W36	Implemented inspection or monitoring program of potential spill or leak sources
W39	Other (Specify in Comments)
	RAW MATERIAL MODIFICATIONS
W41	Increased purity of raw materials
W42	Substituted raw materials
W49	Other (Specify in Comments)
	PROCESS MODIFICATIONS
W51	Instituted closed-loop recycling
W52	Modified equipment, layout, or piping
W53	Changed process catalyst
W54	Instituted better controls on operating conditions (flow rate, temperature, pressure, residence time)
W55	Changed from small volume containers to bulk containers to minimize discarding of empty containers
W58	Other (Specify in Comments)
	Cleaning and Degreasing
W59	Modified stripping/cleaning equipment
W60	Changed to mechanical stripping/cleaning devices (from solvents or other materials)
W61	Changed to aqueous cleaners (from solvents or other materials)
W62	Reduced the number of solvents used, to make waste more amenable to recycling
W63	Modified containment procedures for cleaning units
W64	Improved draining procedures
W65	Redesigned parts racks to reduce dragout
W66	Modified or installed rinse systems
W67	Improved rinse equipment design
W68	Improved rinse equipment operation
W71	Other (Specify in Comments)

Appendix V.5 Activity Codes (Continued)

Code	Waste minimization activity	Code	Waste minimization activity
	Surface preparation and finishing		
W72	Modified spray systems or equipment		
W73	Substituted coating materials used		
W74	Improved application techniques		
W75	Changed from spray to other system		
W78	Other (Specify in Comments)		
PRODUCT MODIFICATIONS			
W81	Changed product specifications		
W82	Modified design or composition		
W83	Modified packaging		
W89	Other (Specify in Comments)		
OTHER SOURCE REDUCTION ACTIVITY			
W99	Specify in Comments		

6 System Approach to Data Analysis

We have learned in Chapter 3 that Subtitle C of the Resource Conservation and Recovery Act (RCRA) establishes a strategy for controlling hazardous waste from generation to ultimate disposal. This management scheme is popularly known as the cradle-to-grave system. Generators and transporters of hazardous waste and owners or operators of treatment, storage, and disposal (TSD) facilities are the links in the cradle-to-grave chain of hazardous waste management.

Subtitle C requires generators to ensure and fully document that the hazardous waste they generate is properly managed. The regulatory requirements for hazardous waste generators include obtaining an identification number from the EPA, packaging and labeling of hazardous waste before transport, manifesting of hazardous waste for transport, record-keeping, and reporting.

The record-keeping and reporting requirements provide the states with a method to track the quantities of waste generated and the movement of hazardous waste. The most important tracking mechanism under these requirements is the biennial report that details the hazardous waste activities during the previous calendar year. We also realize from Chapter 5 that the biennial report requirement is applicable to the hazardous waste generators and the owners or operators of TSD facilities as well. Typically, states house the biennial report in an electronic data base and compile the data systematically.

Although the biennial report serves as the basis for tracking the generation and movement of hazardous waste in the state, waste flow estimations require more than the mere understanding of the operation of the data base management system. At a minimum, accurate estimations require knowledge of federal hazardous waste regulations, state rules, and waste handling practices. This chapter provides a conceptual approach to describe state hazardous waste flow based on applicable rules and regulations.

TYPOLOGY OF HAZARDOUS WASTE GENERATORS

The **generator report,** consisting of the identification and certification (IC), off-site identification (OI), and generation and management (GM) forms of the EPA's biennial reporting system (BRS), provides information on the identity of the generator, the generation level of each type of hazardous waste produced, the destination of the waste, and the transporter used to ship the waste. The submission of the generator report is specified under Title 40 of the Code of Federal Regulations (40 CFR) Part 262. Sections 262.10, 262.41, and 262.44 of the 40 CFR contain provisions that require any business that generates regulated quantities of hazardous waste during a calendar year to comply with the generator reporting requirements. Regulated quantities are determined as 2,200 lb or more of hazardous waste or 2.2 lb or more of acute hazardous waste, generated in any calendar month. 40 CFR 261.33(e) provides a list of acute hazardous waste.

Subtitle C of RCRA requires that all TSDs obtain an operating permit and abide by the treatment, storage, and disposal regulations. A generator who treats or stores hazardous waste on-site for more than 90 days is legally performing the function of a TSD facility. This generator is an on-site TSD and is subject to the TSD requirements. On the other hand, under 40 CFR Part 262, any TSD facility that generates regulated quantities of hazardous waste as a result of its treatment process is also subject to the generator reporting requirements in addition to its obligation to submit the TSD report, which consists of IC, GM, WR, and PS forms of the BRS, to the EPA or the authorized state. The hazardous waste generated by such a facility is known as secondary waste.

TSD facilities receive and handle hazardous waste generated by a variety of generators. A TSD facility that manages hazardous waste generated by other generators under the same ownership is defined as a **captive TSD**. If the senders of hazardous waste are not under the same ownership as the receiving TSD facility, the facility is designated as a **commercial TSD** (U.S. EPA, 1993).[115]

Various regulatory designations on hazardous waste producers result in a functional grouping of waste generators. The four categories of generators are:

1. pure generators, i.e., generators who do not store more than 90 days, treat, or dispose of hazardous waste on-site;
2. generator-TSD, i.e., generators who store, treat, or dispose of hazardous waste on-site;
3. captive TSDs that generate secondary hazardous waste; and
4. commercial TSDs that generate secondary hazardous waste.

This functional grouping is necessary for planning and management purposes because these four categories of generators are all the sources of hazardous waste movement in a state. In order to understand how the four streams interact in a large system, a conceptual approach is provided in the following section.

CONCEPTUAL APPROACH TO HAZARDOUS WASTE FLOW

When planning a state hazardous waste management program, it is imperative that the hazardous waste planner has a good understanding of the patterns of waste flow in the state. The generator report serves as the basis for estimating the flow. However, because of the overlapping functions among hazardous waste generators and TSD facilities, the generator report should be used with caution.

The first stage in understanding a state hazardous waste flow pattern involves separating generators into four classes. The next step is to comprehend how these categories interact in a larger system. In this regard it is necessary to view hazardous waste flow as a set of interrelated activities performed by all categories of generators (Figure 6.1).

Figure 6.1 helps visualize the flow of the hazardous waste from one link to the next link in the chain. More important, it gives an idea of the origin of the waste generation, and it identifies which segments of the flow should be found in the biennial report and which parts represent pure or original hazardous waste generated by the industry.

Figure 6.1 shows that the amount of original hazardous waste generation is the sum of (ao + bo + co + do), which is collectively produced by generators A, B, C, and D. Generator A, who generates ao, is a pure generator who does not have any TSD unit on his business premises. Generator B, who generates bo, has one or a combination of storage, treatment, and disposal units on-site. As a result of the treatment of his original waste, generator B (re)generates secondary hazardous waste in the amount of b1. He may or may not generate nonhazardous waste in addition to b1.

Generator C produces hazardous waste in the amount of co, and generator D in the amount of do. Both generators are owned by the same company that operates a TSD facility for use by C and D. As a result of the treatment of hazardous wastes received from C and D, the captive TSD (re)generates secondary hazardous waste. The TSD is also a generator of hazardous waste, identified as generator E, and the amount produced is eo.

Hazardous waste from individual generators proceed to a commercial TSD for storage, treatment, and final disposition. During the treatment process, the commercial TSD (re)generates secondary hazardous waste. The TSD is also a generator, identified as generator F, and the amount generated is fo.

Figure 6.1 shows that the amount of hazardous waste originally *generated* by pure generators is the sum of (ao + bo + co + do). The quantity *reported* in the biennial report, on the other hand, is the sum of (ao + bo + co + do) + (b1 + eo + fo). This means that there is a potential discrepancy between the reported quantities and the actual tonnage of waste generation.

The difference represents the amount of secondary waste (re)generated from hazardous waste treatment processes. The extent of secondary waste generation can be illustrated by the following statistics. In 1987, the reported quantities of hazardous waste in Arizona may have exceeded the actual amount of hazardous waste generated in the state by as much as 30 percent. Although Arizona is by no means a representative state, there is a national consensus that secondary generation of hazardous waste represents an important issue for hazardous waste planning.

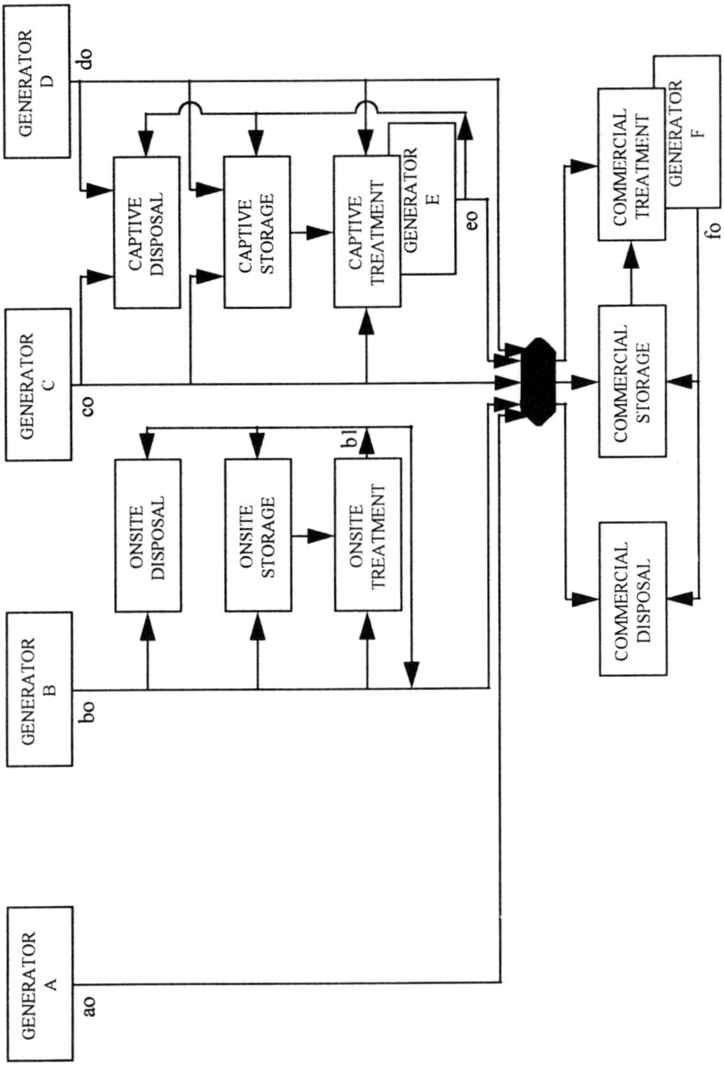

Figure 6.1 Schematic Cradle-to-Grave Hazardous Waste Flow.

Another aspect of reviewing the hazardous waste flow in a state is to examine it from the spatial perspective by dividing the flow into in-state and out-of-state circulations. This can be done by using the generator report of the BRS, because the report provides data on the origin-destination pattern of hazardous waste shipments. The information can be used to examine the portion of the waste that is managed in-state and the amount of waste shipped out-of-state. The import data are retrievable from the **TSD report** of the BRS and can be used to complete the pattern of hazardous waste flow in the state. In the BRS, the TSD report consists mainly of the Waste Received (WR) and Process System (PS) forms. The terms *import* and *export* in the context of the BRS are used to refer to shipments into and out of a state (Figure 6.2).

Figure 6.2 shows that the reported generation of hazardous waste can be examined from two aspects, i.e., from the sources of the report, and the spatial allocation of the generated waste. The sources of the hazardous waste generation report are pure generators A, captive TSD facilities B, and commercial TSD facilities C. They all are in-state hazardous waste handlers.

The reported quantity of hazardous waste generated (box D) can be partially managed in-state (box E) while the remaining quantity can be shipped out-of-state for export (box F). The in-state TSD facilities also import hazardous waste from out-of-state generators (box G). The total hazardous waste managed in-state, box H, is the sum of box E and box G.

Figure 6.2 suggests that the reported hazardous waste generation tonnage may not necessarily be the same as the total quantity of hazardous waste managed in-state. This is illustrated in Figure 6.3 by using the 1987 Arizona data.[116]

Figure 6.3 shows that the total reported waste generated was 28,686 tons. Of this total, 9,138 tons were managed by in-state TSD facilities, and the remaining 19,548 tons were exported to out-of-state facilities. During the reporting period, the in-state TSDs imported 17,688 tons of hazardous waste from out-of-state handlers. This means that the amount of hazardous waste managed in-state totaled 26,826 tons. The following conclusions can be reached from Figure 6.3:

1. The total tonnage of waste managed in-state was not the same as the tonnage of waste generated. In this example, the waste managed was less than the total waste generated.
2. The difference between the amount of waste managed and the amount generated represents the balance of the state's import and export activities. In this example, the waste managed was less than the amount generated because the state was a net exporting state, i.e., the tonnage of exported waste was greater than the imported waste.

Like any other data management system, the BRS also contains certain limitations. Besides the issue of secondary waste discussed in the preceding section, there are some other issues related to the tonnage provided by the generator report.[117] This and other issues related to hazardous waste data management are discussed in the next paragraphs.

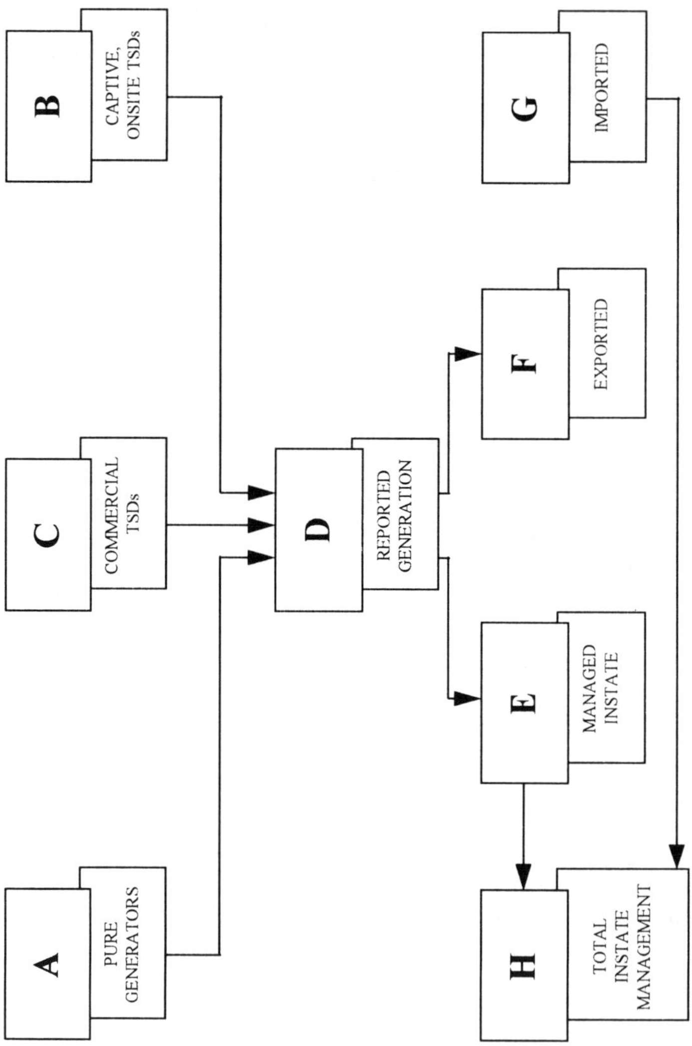

Figure 6.2 Conceptual In-State Hazardous Waste Flow.

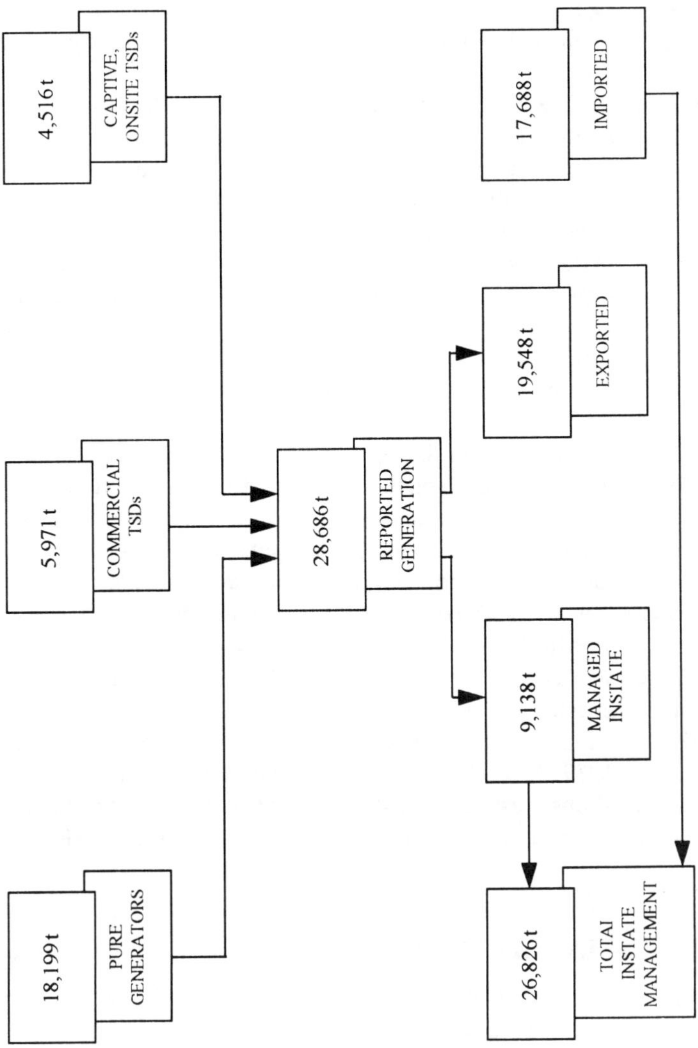

Figure 6.3 Arizona Hazardous Waste Flow, 1987.

DATA MANAGEMENT ISSUES

One of the problems in managing hazardous waste data is *double counting.* Although the secondary waste is normally regenerated by the TSD facilities, it is quite possible that the waste is simply double counted because the facility remanifested the waste for further shipment. This particular case can be addressed by examining the manifest data of individual hazardous waste handlers and comparing the data with information in the BRS submitted by the same handler.

The next issue deals with *exempt reporting* stipulations. As indicated previously, generators that produce less than 2,200 lb of hazardous waste or less than 2.2 lb of acute hazardous waste in any calendar month are exempted from the biennial report. This means that the amount of hazardous waste generated by small-quantity generators (SQGs) is not counted in the BRS format. This problem can be solved by conducting an SQG survey. It is important to scrutinize the necessity of such a survey because of the costs. In Arizona, for example, hazardous waste generated by the SQGs is estimated to be one-eighth of the amount generated by large-quantity generators (LQGs). On the other hand, the number of SQGs is approximately nine times the number of LQGs (Soesilo and Wilson, 1990).[118]

The export-import comparison should also be used with caution, because the accuracy of the comparison depends on the way hazardous waste is defined by the states. The definition of RCRA hazardous waste is covered by 40 CFR Part 261, and most states, including Arizona, normally incorporate by reference the federal regulations into the state rules. Some states, however, have expanded their authority to regulate **state-designated hazardous waste** in addition to the RCRA waste. California is the state with the most elaborate list of state designated hazardous waste, popularly known as California waste. Examples of California listed hazardous waste include asbestos and used oil.

California's file indicates that the state exported 30,118 tons of hazardous waste to Arizona in 1987. Arizona claimed that in that year the state received only 13,273 tons of hazardous waste from California.[119] Clearly, this substantial discrepancy occurs because of the different definitions of hazardous waste between the two states.

DATA PROCESSING ASSISTANCE

Depending on the use of the term, the BRS can provide different interpretations to different people. Some may think of the BRS forms used by generators and facilities to submit the required report to the EPA or authorized states. Others may point to the electronic BRS data system used by the states and the EPA to process the BRS forms. Yet another interpretation is the BRS report based on the processing of the data from the data system.

Currently, the BRS data collection system consists of three tiers, which are presented below (U.S. EPA, 1992):[120]

1. States send BRS forms to and collect from the regulated community, process the data, and send it to the respective EPA regional office.
2. EPA regional offices accumulate states' data, perform data quality checks, and send the data to EPA headquarters.
3. Headquarters accumulates regional data in a national data base and conducts a final data quality check.

The national BRS report, published at EPA headquarters, describes the demographics of, and trends in, hazardous waste generation, management, and minimization. The report is the EPA's information base for the national discussion on the country's hazardous waste problem and the progress achieved in addressing it.

To facilitate the states in developing and maintaining their electronic BRS data system, the EPA provides the states with an ongoing technical assistance program. With this program, the EPA supplies the states with the same basic data system used at EPA headquarters and regional offices.

The data system runs on IBM-compatible PCs and the EPA's IBM mainframe. The system is implemented in FOCUS on both platforms. FOCUS is an informational management and decision support system that provides user-interface tools to guide users through the process of performing various tasks.[121] The system allows the user to key in data from BRS forms into a database, load electronic data into the database directly, perform various data quality checks, run descriptive and management reports, send the data from one system to another, and provide access to the national data for state and federal users.

As part of the EPA's technical assistance program to the states, it issued the BRS User Guide (U.S. EPA, 1992)[122] to assist the BRS data system users in implementing the system. The EPA also provides the BRS User Support hotline, which can be reached by dialing (800) 767-7274. The user support staff will assist the callers with any problems they may have with the BRS software or documentation.

The biennial reporting requirements state that if a facility receives hazardous waste from off-site generators, the facility must submit the WR form of the BRS forms. The WR form collects information about the quantities and characteristics of each hazardous waste type received during the reporting period. For each waste received, the data must be broken down by the off-site generator and also by the treatment or disposal system used to manage the waste. Because a facility may serve a very large number of clients, the combination of the waste type, off-site generator, and hazardous waste management system will result in a significant number of WR forms.

To address this issue, a facility may contact its state environmental agency to find out if electronic reporting is allowed in that state. Electronic reporting is an option where a facility can transmit the report electronically using the BRS data system. If the option is available, the facility has to commit the resources necessary to run the program.

Part Four — Types of Hazardous Waste Planning

Hazardous waste planning creates plans, policies, or programs. Depending on the specific planning objectives, the activity can be generalized into four major groups. They include (1) hazardous waste management planning, (2) site correction and remedial planning, (3) emergency response and hazard management planning, and (4) citizen participation planning.

Chapter 7 discusses hazardous waste management planning, which can be described as an activity that supports the goal of hazardous waste management in assuring the proper management of hazardous waste from the time it is generated to its ultimate disposal. In the context of hazardous waste management, compliance planning represents the first planning category, followed by waste minimization and pollution prevention, capacity assurance planning, facility siting, and state hazardous waste management planning. These five categories of planning are the topics of Chapter 7.

Chapter 8 describes the planning aspect of site correction and remediation. Based on its regulatory status, a hazardous waste contaminated site may be addressed from three alternative approaches, i.e., as a CERCLA site, an RCRA corrective action site, or an RCRA contaminated site. These three categories of sites are discussed in Chapter 8.

Chapter 9 is about emergency response and hazard management planning. Emergency response and hazard management planning are concerned with the safety of people and the environment from potential hazards resulting from hazardous materials accidents, incidents, or improper management of hazardous material. The planning aspect of the hazard management process is addressed in this chapter.

Chapter 10 examines citizen participation planning. Citizen participation planning is essentially about building and maintaining effective and efficient communication among all of the parties involved in the hazardous waste management system: the business, the government, and the public. Citizen participation planning in both the private and public sectors is described in this last chapter of Part Four.

7 Hazardous Waste Management Planning

Hazardous waste management planning is an activity that supports the goal of hazardous waste management. The goal is to assure the proper management of hazardous waste from the time it is generated to its ultimate disposal. To foster high levels of hazardous waste management compliance, the EPA and authorized states utilize a combination of enforcement approaches that promote voluntary compliance by the regulated community and that intensify compliance monitoring activities — including inspections to detect violations — and take firm but fair enforcement action against violators to correct violations and create a strong enforcement presence (U.S. EPA, 1993).[123]

In the 1990s, one of the most notable propensities in the domain of hazardous waste management is the increase of the government's effort in its environmental crime-prevention enforcement activities, which force more companies to develop sensible hazardous waste compliance programs in order to avoid prosecution. When it comes to non-compliance, companies know they are facing a potentially large monetary fine; and if they knowingly violate hazardous waste laws, they realize that criminal prosecution is a reality.

Hopkins (1993)[124] reveals that regulatory compliance currently represents a hot issue in the industrial community. One of the problems in addressing compliance is that some companies are facing obstacles in developing a good compliance program. The obstacles include limited resources, a lack of information on regulatory requirements, and uninformed and untrained environmental personnel. Because of these obstacles, many small to mid-size companies say they are not achieving the compliance they should.

While larger corporations usually have environmental planners, lawyers, consultants, or established programs to help satisfy compliance needs, smaller and mid-size companies normally do not have that level of expertise or support. The companies that have no compliance programs and are recipients of penalties for violations of environmental laws may find that not having a compliance program is more expensive than having one.

In the field of hazardous waste management, compliance planning includes a framework for understanding the complex web of regulatory requirements and a guide for waste management decision-making within the regulatory context

(Lindgren, 1989).[125] This regulatory framework and compliance decision-making guidance is extensively elaborated by Lindgren in his "Managing Industrial Hazardous Waste" handbook.

The scope of hazardous waste management planning obviously extends beyond regulatory compliance planning. Waste minimization and reduction of sources of waste generation, for one thing, are also important components in any hazardous waste management planning. Other types of planning activities include capacity assurance planning, facility siting, and state hazardous waste management planning. These five planning types are the topics of discussion of this chapter.

REGULATORY COMPLIANCE PROGRAM

Overall, the purpose of a regulatory compliance program is to ensure that the facility is in full compliance with all hazardous waste regulations. The most logical place to start with a regulatory compliance program is to determine if you generate hazardous waste, and if so, to what extent. This determination can be made by taking an inventory of each waste stream generated from the facility. The term "waste stream" refers to waste materials generated from a known point of origin. Examples of waste streams are spent solvents from vapor degreasers, paint booth sludges from water wall paint booths, and wastewater treatment sludges from electroplating operations. Once each waste stream is identified, then determine whether the waste is hazardous within the regulatory context. In most cases, a facility that generates hazardous waste will also temporarily store waste, and perhaps treat some waste on-site, before the facility ships it to off-site disposal facilities.

With this generate-store-treat scenario as the basis of discussion, the general regulatory inspection protocol normally focuses on the following components (U.S. EPA, 1992):[126]

1. Examine points of waste generation and waste accumulation. The purpose is to determine if all hazardous wastes have been identified and if accumulation points meet hazardous waste satellite storage area regulations.
2. Examine in-plant waste transport from generation points and accumulation points to storage and treatment units. Ensure that there are no violations for labeling, placing, or handling waste, and determine that the wastes are adequately tracked to enable proper identification at storage and treatment units.
3. Examine storage and treatment units for compliance with applicable standards. Ensure that wastes in units correspond to those whose points of generation have been evaluated. If not, identify where any other wastes in the units come from, and determine if they are hazardous wastes or not. In addition, identify any hazardous wastes generated as a result of treatment process, e.g., treatment sludge, and assess the management of this secondary waste for compliance.

Cahill and Kane (1992)[127] present a list of typical compliance problems found in a survey of the results of hazardous waste audits. An audit, as defined by the EPA, is a systematic, documented, periodic, and objective review by regulated entities of facility operations and practices related to meeting environmental requirements (U.S. EPA, 1986).[128]

Cahill and Kane's list basically converges around three major areas of frequent violations. They are:

1. inadequate waste determination resulting in hazardous waste being handled as nonhazardous waste,
2. incomplete hazardous waste manifests, and
3. problems related to storage of hazardous waste, such as:

   - Accumulation point was not maintained with proper segregation of incompatible wastes.
   - Hazardous waste containers were stored more than 90 days.
   - Hazardous waste containers did not have information on accumulation start dates, and other necessary information was incomplete.
   - Open top funnels and bungs were missing on hazardous waste drums.
   - Grounding clips and wires were missing on highly flammable hazardous waste drums in storage.
   - Satellite accumulation drums were not properly labeled.
   - Quarterly inspection reports of accumulation points were incomplete or not available.
   - Contingency and emergency response plan was not available or maintained at the accumulation point.
   - Emergency response equipment and materials were not available or were inadequate at the accumulation point.
   - The manager of the accumulation point was not properly trained and was not specifically designated in writing.

The regulatory inspection protocol components and typical compliance problems discussed above provide clear indications as to the direction and target areas of a hazardous waste compliance program. Because full compliance is desired, compliance assurance can best be achieved by reviewing:

1. current practices of hazardous waste determination,
2. generator status identification,
3. satellite accumulation procedures, and
4. recycling exemptions related to the facility.

The purpose of these reviews is to find regulatory areas that are normally vulnerable to noncompliance.

The focus of the reviews of those four **target areas** are:

1. *To take inventory and identify each waste generated from the facility.* This includes the identification of secondary waste generation and, if applicable, the wastes that are stored, treated, or disposed by waste management type. Due to periodic cleaning of equipment, discovery of expired chemicals, and discovery of materials contaminated with hazardous waste due to spills or leaks, the facility may also generate one-time hazardous wastes. Once identified, determine whether each waste is a hazardous waste within the regulatory framework (RCRA). It is also important to check whether the existing determination is correct by focusing on wastes that are not currently classified as hazardous wastes. Keep in mind that there may be changes in the company's production processes that generate different wastes or changes in the regulatory definitions of hazardous waste at either the federal or state level.
2. *To quantify the total amount of hazardous waste generated per month.* This quantification is needed for determining the generator status of a facility, whether it is a large-quantity generator (LQG), small-quantity generator (SQG), or conditionally exempt SQG. The facility's regulatory status is not static or permanent. The status can go up or down. This year's LQG may have had an SQG status last year, and next year the same facility may be back to its SQG status. Make sure that the generator status determination is correct, because the burden of proof with respect to the applicability of any reduced requirements is placed on the generator.
3. *To examine how each waste is stored.* An LQG may legally accumulate hazardous waste on-site for 90 days or less without a storage permit, provided that the accumulation is properly done. This means that the waste is appropriately placed in containers, tanks, or on drip pads; the date upon which each period of accumulation begins is clearly marked on each container; and, while being accumulated on-site, each container and tank is clearly marked with the words "hazardous waste." In addition, the facility must be maintained and operated to minimize the possibility of a fire, explosion, or any release of hazardous waste into the environment, through proper equipment (alarm system, communication system, fire extinguishers, and sprinklers); proper testing and maintenance of equipment; easy access to alarm and communication systems; and sufficient aisle space. The facility must also have a contingency plan in case of an emergency, and it must have prior arrangements with local police departments, fire departments, and emergency responders in regard to emergency prevention and preparedness.
4. *To examine how each waste is managed.* A generator that treats or disposes of hazardous waste on-site is subject to, respectively, a treatment permit or a disposal permit. RCRA regulations do not consider recycling as a regulated treatment process; therefore, recycling is exempted from the hazardous waste treatment permit requirement. Generally speaking, hazardous wastes destined for recycling are subject to,

as applicable, regulations for generators, for transporters, and probably for storage facility requirements. This means that if a recycler/generator stores hazardous waste for more than 90 days, as a generator the recycler is required to obtain a hazardous waste storage permit. In addition, because of the recycling exemptions, it is necessary to ensure that the recycling activity in the facility, if any, fits the regulatory definition of recycling. If the activity does not fit the definition, the facility may be required to obtain a hazardous waste treatment permit.

The compliance targeting discussed above is intended as general guidelines and serves to illustrate a typical compliance planning scenario. The scenario will obviously vary from one company to another depending on the size and nature of the company. With those four components as target areas, a detailed schedule of follow-up activities can be developed. The schedule represents a variety of efforts including document reviews, site visits, staff interviews, data analysis, and plan formulation. The conclusive result of these activities will be the development of a final compliance plan.

Compliance Planning

Compliance planning translates the four compliance target areas, discussed in the preceding paragraphs, into a set of activities, or tasks, that are necessary to accomplish the ultimate compliance objectives. These tasks are:

1. Review chemical purchasing records. Because most hazardous wastes are generated from chemicals, the fate of each chemical can be examined by tracing its path through the plant processes or operation. From this review, a list of chemicals that are potentially responsible for hazardous waste generation can be developed. Cleaning agents, process chemicals, industrial oils, water treatment chemicals, and surface finishing agents are usually included in the list.
2. Contact the vendors selling the potentially responsible chemicals and request them to furnish Material Safety Data Sheets (MSDS). The purpose is to obtain preliminary information on the chemical composition of the chemicals. Such information is important when deciding whether the chemicals used meet any hazardous waste "listing" constituents.
3. Review hazardous waste manifests and flow charts or diagrams showing processes at the facility. The purpose of the review is to obtain a basic understanding of the relationship between input chemicals and waste materials.
4. Perform site visits and conduct interviews with equipment operators, maintenance crews, and other housekeeping personnel. This effort is aimed at tracking where, how, and why portions of the input chemicals are becoming waste materials. During these visits, it is necessary to

examine the facility's routine operations and identify any noncomplying procedures that are part of the operations. Also, examine unusual situations that deviate from the facility's normal operating procedures that may create potential for regulatory violations.

5. Draft an initial input-output diagram showing the relationship between the input chemicals, various operations and processes, and the waste materials generated.
6. Quantify each waste stream by using historical data and, if necessary, by utilizing appropriate sampling methods.
7. Determine which waste stream can be categorized as regulated hazardous waste. If necessary, use appropriate laboratory analyses on representative samples to assist in making the determination.
8. Total the quantity of hazardous waste generated, and determine the generator status of the facility.
9. Determine the legality and appropriateness of existing storage practices and final disposition of all regulated waste materials. During this process, it is also important to examine steps in the waste handling practice during which wastes may have been mislabeled and mishandled and, therefore, created potential for spills or releases.
10. Review the record-keeping and reporting documentation, including the manifest, the biennial report, and other reports submitted by the facility to the EPA or to the state agency: consultants' reports, permit documents, and laboratory reports.
11. Examine the facility's compliance history, including past violations cited in inspection reports from the EPA or the state agency, efforts to correct compliance problems and potential violations which may not have been corrected as documented in copies of compliance correspondence with the EPA or the state agency, and records of phone conversations with officials.
12. Document the findings of all analyses and reviews, and draft a compliance plan.

Once upper management approves the draft, the final compliance plan can be formalized. Generally, a plan developed using this process represents a retroactive mechanism to compliance. By expanding the scope of a compliance plan, a proactive compliance effort can be accomplished. The next section will address this proactive approach to compliance.

Compliance Strategy

A compliance strategy is a strategic plan for achieving an effective compliance program. The scope is broader than a compliance plan. The basic philosophy of a well-planned compliance strategy is that the best defense to an enforcement action is active and aggressive compliance. A compliance program will be truly effective only if self-imposed compliance standards are followed and violations are corrected promptly upon discovery (Green, Dunn, and Oppenheimer, 1993).[129]

Green et al. suggest a number of strategic initiatives. The following components are among the initiatives that need to be included in the compliance strategy:

1. A written compliance policy. Ensure that everybody in the company subscribes to it.
2. A training plan consisting of regular training and refresher courses. The goal is to ensure that each employee understands what the law requires.
3. A written procedure to deal with regulators, particularly with state or EPA inspectors.
4. An internal system that is able to discover and accommodate employee complaints. Many "informants" are usually company employees frustrated by the lack of response to complaints by higher-ups.
5. A contingency plan. This should be in place in case of criminal investigations.
6. A prudent record-keeping policy.
7. A well-established environmental information and reporting system.
8. A protocol to monitor and address violations. Identify and correct violations promptly; do not ignore small violations, because they often snowball.

Monitoring violations, as part of the compliance strategy, can be accomplished through an internal environmental audit. An environmental audit is a preventive measure that can improve facility performance before the state or the EPA inspector arrives. It can identify environmental management deficiencies; however, the audit cannot correct the deficiencies (Kuusinen, 1992).[130] Additionally, an audit report noting a problem that is not later addressed by a facility can be used by prosecutors to prove "knowing" or "negligent" criminal conduct against the facility (Shanoff, 1993).[131] Companies may take a variety of legal steps to protect audit reports; however, even the best-laid efforts at protection may be overruled by a judge ordering disclosure in specific circumstances (Levin, Hymes, and Mullaney, 1994).[132] Audit reports should always be drafted with the possibility of disclosure in mind.

Compliance planning is consistently challenging to planners, the hazardous waste regulatory program is always in a state of flux. The promulgation of new regulations and modifications to and changing interpretations of regulatory provisions make the management of a compliance program demanding. On the economic side, there are few benefits because of the increasingly high costs of pursuing only regulatory compliance (Hirschborn, 1991).[133] Additionally, compliance efforts with respect to one environmental medium (air, water, or land) may adversely affect compliance efforts in another.

A well-executed compliance strategy will certainly provide a company with an effective hazardous waste compliance program; however, the program will not necessarily result in efficient hazardous waste management. To achieve efficiency and effectiveness in managing its hazardous waste, a company is required, among other things, to minimize the generation and toxicity of its hazardous waste to the extent feasible. This implies that the company should implement some sort of waste minimization program.

WASTE MINIMIZATION AND POLLUTION PREVENTION

In this decade and beyond, having a well-planned waste minimization program in place is a necessity for businesses to stay ahead of the competition. In the private sector, waste minimization planning has become a more and more important aspect of business programs than it was in the past. Similarly, the importance of state waste minimization programs, or comparable programs at the local government jurisdiction level, is also recognized by the public sector. This section describes waste minimization planning in both the private and public sectors.

The term *waste minimization* came into use in 1976 with the passage of RCRA; since then, it has been heard increasingly at various meetings of those working in the field of hazardous waste planning and management. Waste minimization is focused on hazardous waste and is defined as the reduction — to the extent feasible — of hazardous waste that is generated or subsequently treated, stored, or disposed of. It includes any source reduction or recycling activity undertaken by a generator that results in either (1) the reduction of the total volume or quantity of hazardous waste, or (2) the reduction of the toxicity of the hazardous waste, or both, as long as such reduction is consistent with the goal of minimizing present and future threats to human health and the environment (U.S. EPA, 1986).[134]

Pollution prevention is a term closely related to waste minimization, and it came into use in 1990 with the passage of the Pollution Prevention Act. Pollution prevention, however, is focused on all forms of pollution, not just hazardous waste, and includes source reduction, in-process recycling, and reuse activities that eliminate or reduce potential or actual releases of pollutants to the environment. Some states, such as Arizona, Illinois, Indiana, Maine, Massachusetts, Minnesota, New Jersey, Oregon, Vermont, and Washington, have adopted a broader definition of pollution prevention, and it includes all forms of recycling, some forms of treatment, and the reduction of the use of toxic substances (U.S. GAO, 1992).[135]

With both pollution prevention and waste minimization, there is a preference for source reduction over recycling and recycling over treatment from an environmental perspective. The preference is based on a national environmental management hierarchy, which establishes a priority scheme from the highest to the lowest of the following management options: the first priority is source reduction, the second is recycling, the third is treatment, and the fourth — the least preferred — is disposal. This priority scheme is also known as the hazardous waste management hierarchy.

Source reduction is defined as any activity that reduces or eliminates the generation of hazardous waste or pollution at the source, usually within a process. Source reduction activities include changes in input material, changes in technology, product changes, and good operating practices such as improvement in material handling, segregating waste streams, preventing chemical spill, and improvements in production scheduling.

A waste or pollution is *recycled* if it is (1) used or reused, or (2) reclaimed. It is used or reused if the waste or pollution is returned to the original process or used as a raw material substitute for another process. A waste or pollutant is reclaimed if it is processed for resource recovery or processed as a by-product.

Treatment means any method, technique, or process, including neutralization, designed to change the physical, chemical, or biological character or composition of any hazardous waste. The purpose is to neutralize the treated waste, to recover energy or material resources from the waste, to reduce the volume, or to render such waste less hazardous or nonhazardous, so that the waste is safer for storage, transportation, and disposal.

Because pollution prevention addresses the release of all types of pollutants to all three environmental media, the transfer of pollutants from one medium to another is not pollution prevention. Although the focus of waste minimization is only on hazardous wastes, it is important that all pollutant emissions into air, water, and land resulting from the management of a hazardous waste is also considered in a waste minimization program. Therefore, similarly, the transfer of pollutants from one environmental medium to another is not waste minimization. In fact, waste minimization should help ensure that the reduction of pollutants in one medium does not shift or create pollutants or risk in another medium. In general, the variety of techniques used to reduce the releases of contaminants in a pollution prevention program are similar to the techniques used to reduce the generation of hazardous waste as applied to the waste minimization program. Because of this, the terms *pollution prevention* and *waste minimization* in this section are used interchangeably.

An example of cross-media shifts is well described by Anderson and Herb (1992).[136] They illustrate that in order to remove volatile organics from stack emissions, a facility installed a wet air scrubber. The effort simply resulted in the organics being transferred from the air into the water, because water from the scrubber must be discharged into a nearby stream. The discharge required a modification of the facility's water discharge permit, which eventually forced the facility to install an air stripper, transferring the organics back into the air. Both pollution prevention and waste minimization should strive to prevent this type of cross-media transfer.

As of 1992, 23 states had some pollution prevention, waste minimization, or source reduction planning legislation or requirements (Foecke and Style, 1992[137]; OEPA, 1991[138]). In general, most states require a pollution prevention or waste minimization plan from certain industries, typically large-quantity generators of hazardous waste or Form R filers. Table 7.1 provides a list of those states and includes the filing status of the plans and the plan progress report requirements of each state.

The table shows that of the 23 states, 16 require that the regulated businesses complete a pollution prevention plan, while 7 states allow businesses to complete a plan on a voluntary basis. Fifteen of the states listed also require the submittal of a plan progress report that describes the results of implementing pollution

Table 7.1 States Requiring Pollution Prevention, Source Reduction, and Waste Minimization Planning

No.	States	Filing status	Progress report
1	Arizona	Mandatory	Yes
2	California	Mandatory	Yes
3	Connecticut	Voluntary	No
4	Delaware	Voluntary	No
5	Florida	Voluntary	No
6	Georgia	Mandatory	Yes
7	Illinois	Mandatory	No
8	Indiana	Voluntary	No
9	Iowa	Voluntary	No
10	Louisiana	Voluntary	No
11	Maine	Mandatory	Yes
12	Massachusetts	Mandatory	Yes
13	Minnesota	Mandatory	Yes
14	Mississippi	Mandatory	Yes
15	New Jersey	Mandatory	Yes
16	New York	Mandatory	Yes
17	North Carolina	Mandatory	Yes
18	Oregon	Mandatory	Yes
19	Tennessee	Mandatory	Yes
20	Texas	Mandatory	Yes
21	Vermont	Mandatory	Yes
22	Washington	Mandatory	Yes
23	Wisconsin	Voluntary	No

prevention. The list is obviously changing over time and will most likely expand in the future. This means that pollution prevention and waste minimization is becoming a high priority for businesses in order to stay competitive.

While the national environmental management hierarchy ranks source reduction as the first management priority, land disposal is considered the least preferred management practice because of its considerable adverse impact on the environment. As such, numerous regulations and restrictions have been established to strongly *discourage* this practice, the primary one being the land disposal restrictions (LDR) established through the Hazardous and Solid Waste Amendments (HSWA) of 1984. Because these LDRs have increased the importance of waste minimization, it is relevant at this point to briefly examine the requirements found in the land disposal restrictions.

Land Disposal Restrictions

The LDR regulations established treatment standards for the majority of RCRA hazardous wastes ultimately destined for land disposal (U.S. EPA, 1992).[139]

The regulations differentiate this majority into *restricted* and *prohibited* wastes. Restricted wastes are RCRA wastes for which an LDR treatment standard has been established — for example, D004, D005, D006, D007, D008, D009, D010, and D011 wastes. The standard is expressed as concentration levels or as methods of treatment that have either substantially diminished the toxicity of wastes or reduced the likelihood that hazardous constituents from wastes will migrate from the disposal site (U.S. EPA, 1991).[140] Before the regulated wastes can be land-disposed, LDR treatment standards must be met. Prohibited wastes, which constitute a subset of restricted wastes, are ineligible for land disposal. Examples of prohibited wastes are solvent wastes and dioxin-containing wastes.

The EPA has promulgated regulations implementing the LDR program according to a specified rule-making schedule. LDR regulations in November 1986 included prohibitions on the land disposal of dioxin and solvent containing hazardous wastes. The July 1987 regulations banned the land disposal of certain hazardous wastes which California had already banned.[141] With the LDR program fully effective since May 1992, a significant quantity of RCRA wastes have to be properly treated prior to land disposal. This results in higher costs in managing hazardous waste, because treatment costs are expensive. As a matter of fact, treatment costs are considered higher than disposal costs.

It is obvious from the above discussions that Congress wanted to promote waste minimization and pollution prevention in lieu of, or prior to, land disposal (Case, 1993).[142] A recent study by Graham (1993)[143] on pollution prevention practices reveals that the two driving factors frequently cited by the industry in implementing waste minimization and pollution prevention programs are compliance with the law and cost savings from disposal and treatment. Unquestionably, motivation to implement pollution prevention has always been the key to a successful program. The next section discusses how the program should be initiated, developed, and implemented once the motivation has been established.

Pollution Prevention Planning in the Private Sector

Pollution prevention represents a comprehensive and continual planning effort. Before a company becomes committed to the program, the management must be convinced that pollution prevention is technically and economically feasible. The company must realize that the benefits of such a program will outweigh its costs.

A large body of literature has been devoted to raise the level of awareness about pollution prevention and to encourage businesses to implement pollution prevention (INFORM, 1992).[144] It is widely known that the benefits of implementing pollution prevention include:

1. reduced handling/treatment/disposal costs, raw material purchases, and other operating costs;
2. reduced insurance costs and risk of long-term and short-term criminal and civil liabilities;

3. strengthened protection of worker safety, public health, and the environment;
4. improved employee morale and participation;
5. improved product quality;
6. reduced regulatory burden; and
7. enhanced company image.

While the benefits are recognized, the limitations of pollution prevention are also considered. Graham's 1993 survey indicates that the following factors represent impediments to a successful pollution prevention program:

1. high cost of new equipment;
2. weak management support;
3. lack of time, and staff shortages;
4. lack of appropriate technologies;
5. difficulty in meeting customer quality standards; and
6. poor accounting systems to quantify the costs of wastes and disposal.

Even with these barriers confronting the industry, pollution prevention is still desirable and highly feasible. The industry keeps learning from success stories as well as from less successful efforts of others. These stories are useful to prevent future mistakes.

There are five major planning steps in implementing a pollution prevention program. They are (1) establishing the program, (2) assessing pollution prevention opportunities, (3) writing the plan, (4) communicating the plan, and (5) implementing and monitoring the plan. The following paragraphs briefly outline the activities involved in each step.

Establishing the Program

The pollution prevention initiative normally comes from top management. A successful program, however, requires not only a commitment by management but also an awareness among employees and proper planning. Although there is no standard way to begin a program, there are usually three components that serve as a starting point for a business. The components are:

1. Establish a formal written policy stating management's commitment to preventing pollution.
2. Appoint a program coordinator and champion of the program, and establish a company pollution prevention team.
3. Set pollution prevention goals.

Once the first component is developed, the *pollution prevention policy* should be distributed to all employees to create an awareness of pollution prevention. While management will set the tone for the program, the attitude of employees will have a significant impact on the program's success. The policy statement is an important initial step, regardless of the size of the company.

The statement should express why a pollution prevention program is being established, what is to be accomplished, and who will do it. Policies differ among companies, but they usually include the following statements:

- That the company wants to protect the environment, will reduce or eliminate the amounts of all types of waste or emissions, and will improve energy efficiency.
- That everyone in the company will be involved.

The language of the policy statement varies from one company to another. A short version of the statement, from the *Minnesota Guide to Pollution Prevention Planning* (MOWM, 1991),[145] is provided below as an example:

> At (the company name), protecting the environment is our high priority. We pledge to eliminate or reduce our use of toxic substances and our generation of all wastes or emissions, whenever possible. Prevention of pollution at the source is the preferred alternative. When wastes or emissions cannot be avoided, we are committed to recycling, treatment, and disposal in ways that minimize any undesirable effects on air, water, and land.

After the establishment of the policy, management must support the policy and be willing to commit the resources to sustain the program. At small firms the commitment may be obvious. At large companies with complex management systems, a concise statement of this commitment is essential — and management must focus and strongly participate in the process. Clear responsibility for the pollution prevention effort should be established; and, depending on the size of the company, *a person or a team should be appointed* for developing and implementing the program.

The team may consist of personnel from a variety of departments within the company in order to effectively deal with the wide scope of pollution prevention. The team can include, for example, staff from the purchasing department, manufacturing, facility operation, material engineering, engineering laboratory, and management. The person or the team assigned by the management should start planning the undertaking by concentrating the company's program on specific goals.

The setting of *pollution prevention goals* should be given serious consideration, because well-defined goals will help to focus effort and build consensus. A company may adopt a corporate-wide goal of 10 percent waste reduction per year or consider starting the goal of 100 percent reduction. The last option is probably not a completely achievable goal; however, zero-discharge goals encourage an attitude of continually striving for improvement (U.S. EPA, 1992).[146]

Once the policy is developed, the team created, and the goals established, then the pollution prevention program has been officially established. The next step is to assess pollution prevention opportunities available in the facility.

Assessing Pollution Prevention Opportunities

Pollution prevention opportunities are the changes to be made or activities to be implemented that will result in reduced emissions or waste generation. Because pollution prevention cannot be accomplished on all of a facility's operations, waste streams, or pollutant sources at once, it is necessary to target specific areas for the prevention effort. In order to do that, the team must first assess the overall operations and processes, products and services, and waste streams or toxic chemical usage at the facility, then prioritize and select certain target areas. This analytical process or assessment is perhaps the most important part of the overall pollution prevention planning effort (ADEQ, 1992).[147]

To simplify the discussions, the scope of assessment can be best conceptualized by the process flow diagram shown in Figure 7.1. The diagram rudimentarily illustrates the relationships between chemical usage, operations and processes, products and services, and waste streams.

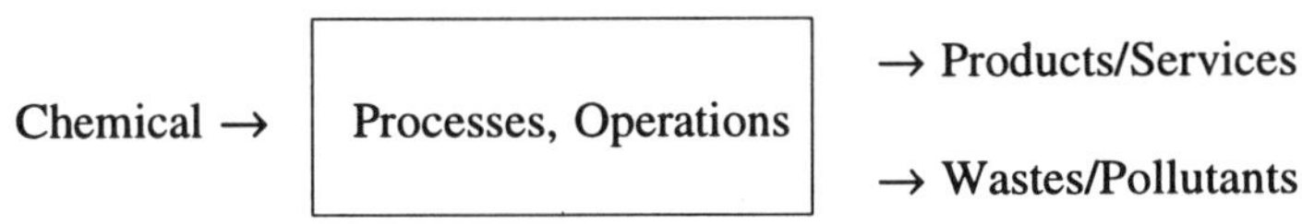

Figure 7.1 Process Flow Diagram.

In general, the assessment consists of a review of the four components shown in Figure 7.1 and will involve the collection of background information such as flow charts, diagrams, drawings, sketches, specifications, material balances, and other information that will clearly illustrate the overall flow of materials used, products produced, wastes generated, and toxics released. To assist planners with performing this analytical evaluation, the EPA in 1992 published a manual entitled "Facility Pollution Prevention Guide," which provides the readers with comprehensive assessment procedures. The manual is furnished with a series of sample worksheets that can be used to assist in collecting, compiling, and displaying data (see Appendix VII.1).

The summary of the procedure is described below:

1. Identify the major **products or services** produced by the facility. Examples of the type of major products or services, and the respective units of measure, are:

    printed circuit board → number of square feet
    paint → number of gallons
    auto repair shop → number of car engines repaired

    Additional information that is required includes the quantity of products produced, sold, and reworked; the average production cost per unit; the average sale price per unit; the annual revenues; the on-site storage mode (warehouse, outdoors, underground tank); the shipping mode (truck, pipeline, railway tank car), and the shipping container type and size (55-gal drum, 100-lb paper bag). The sources for this information are production records, sales receipts, and cost accounting documents.

2. For each product or service, define the **operations and processes** used, and, if possible, break the activities down into individual steps. For each step, evaluate if the following information is currently available, complete, and up-to-date, and specify where to locate the information: process flow diagrams, flow measurements, operating procedures and manuals, process descriptions, equipment lists and specifications, facility or equipment layout drawings, operator logs, production schedules, and inventory records.
3. For each operation or process, identify the **chemicals used,** and describe how much is used, the primary function (feedstock, catalysts, intermediary), and its fate (consumed in products, reused, or discarded as process waste). Additional necessary information includes the name of the chemical (e.g., sulfuric acid, methyl ethyl ketone), the name of the supplier, the physical and chemical characteristics, the annual consumption rate, the annual cost, the shipping mode, the shipping container type and size, the storage mode, the transfer mode (forklift, conveyor, pump), and how the company manages empty containers (return to supplier, recycle, landfill). The sources of this information are chemical inventories and purchasing records, Form R, operating manuals and process descriptions, production records, MSDS, and operator data logs.
4. For each process or operation (including maintenance activities), identify the **waste streams or pollutants** generated. For each waste stream or pollutant (e.g., spent cleaning solvent, chemical baths), describe the quantities of waste generated, the physical or chemical composition, the waste collection method, the type and size of container, the treatment and disposal methods, and the disposal costs. Additional information that will be useful for appropriate assessment includes: regulatory compliance requirements, potential liability of mishaps, safety hazards, source reduction potential, on-site recycling potential, and by-product recovery potential. Updated information from the compliance plan will be useful for this assessment. If that is not available, existing data can be obtained from the facility's production records, permits, manifests, biennial reports, MSDS, Form R, laboratory reports, process specifications, chemical inventory, purchasing records, raw material usage records, wastewater flow records, NPDES monitoring reports, and internal audits.

Once all the background information has been collected for those four analytical areas, it is necessary to put the information all together as illustrated by the process flow diagram displayed in Figure 7.1. Information compiled in this form will serve to illustrate the overall flow of materials at a facility in accordance with the material balance rule. In its simplest form, a material balance concept states that the amount of energy or material that goes into a process will be the same as the amount that leaves the process.

Congruent to this concept, the total mass of all raw materials going into the process must be equal to the output or the total mass leaving the process. This output includes all identifiable products, by-products, solid wastes, hazardous wastes, wastewater, and air emission. The accuracy of this input-output matching exercise will vary depending on the nature and size of a facility. Sufficiency of information is necessary in order to perform an adequate mass balance analysis. At this point, the team should be able to identify any data gaps, data inconsistencies, or missing information, and then address and correct these deficiencies.

Based on the updated background information, the team can then move forward to *prioritize target areas* for pollution prevention. The targeted areas can be a particular waste stream (e.g., heavy metal wastes), chemical used (e.g., solvent), process and operation (e.g., electroplating), or any combination of the above. As a practical tip, Cahill and Kane (1992)[148] suggest that the team investigates the following target areas first:

1. The "devil's" chemicals. In most industry, there will be a few chemicals that are extremely toxic and/or troublesome.
2. High-volume waste streams. The 90/10 rule often applies here; i.e., 90 percent of the waste volume is generated by 10 percent of the waste streams.
3. Highly regulated wastes. They include dioxins, chlorinated solvents, and acutely hazardous wastes, particularly when they have become "political" wastes.
4. Land-disposed wastes. The "strict, joint, and several" liability tenets suggest that the generator of these wastes still maintains some liability.

Once prioritization of target areas is completed, the next step is for the team to become as familiar with each area as possible by conducting a site review and by interviewing employees. The goal of this activity is to obtain more comprehensive information of the targeted areas in order to identify pollution prevention techniques appropriate for the areas.

For each targeted area, the team should evaluate the applicability and effectiveness of the following pollution prevention opportunities:

A. *Source Reduction Techniques:*

1. input changes or material substitutions
2. changes in technology or production process
3. product redesign or reformulation
4. material handling, production scheduling, and other good housekeeping

B. *Recycling Techniques:*

1. reclamation and in-process recycling
2. reuse

Regardless of the type of company, the following pollution prevention tips represent “common sense” techniques that are applicable to various processes and operations:

1. Use non-chlorinated solvents or aqueous cleaners for cleaning.
2. Use water-based coatings and paints.
3. Use oil-water separators in sewer systems.
4. Set vessel run schedules to minimize the need for frequent clean-outs.
5. Implement first-in, first-out inventory control.
6. Avoid cross-contamination of used oil with solvents.
7. Set internal clearinghouse function for excess inventory, old chemicals, and recoverable wastes.
8. Keep lids on degreasers.
9. Avoid spills during materials transfer and dispensing.
10. For research and development, purchase chemicals in small amounts.
11. Require secondary containment at all fill and dispensing areas.
12. Segregate chemical storage from waste storage.
13. Isolate solid wastes from hazardous wastes.
14. Practice preventive equipment maintenance.

The applicability of these techniques to each potential target area must be thoroughly evaluated in order to develop *a list of specific potential pollution prevention opportunities.* A number of measures can be used to compare pollution prevention opportunities and to select the most effective instruments. The measures include the following:

1. quantity of wastes that can be reduced;
2. amount of dollars that can be saved;
3. current and future compliance with regulatory requirements;
4. cost of implementation and pay-back;
5. ease of implementation and training needs; and
6. impact of the opportunity on other activities, such as better material management, optimization potential, and opportunity for quality improvement.

Consider the following questions as screening tools:

1. What are the main benefits to be gained by implementing the project?
2. Does the technology exist to implement the project?
3. Can the project be completed within a reasonable amount of time?
4. Does the project have a good “track record” (i.e., successful implementation by others)?
5. Is the project compatible with the existing facility operation?
6. How much down time and training will be required?
7. How much does it cost, and does it appear to be cost-effective?

Once the potential opportunity has passed the screening, it needs to be further evaluated to determine if it is, in fact, feasible. In general, a thorough economic, environmental, and technical feasibility study for each potential opportunity should be conducted. The output of this process is a list of *feasible pollution prevention opportunities.* Although a potential project may be feasible, it does not mean that it must be implemented. There may be a number of pollution prevention opportunities feasible for a particular waste stream; however, only one will probably be implemented. Normally it is not advisable to implement all feasible opportunities at once.

The successive evaluation procedure in pollution prevention program planning, i.e., from the selection of potential targeted areas (waste, process, or chemical as targets) to the identification of potential pollution prevention opportunities and the determination of feasible pollution prevention opportunities, reflects the complexity of pollution prevention planning. Once the evaluation is completed, it is the responsibility of the team to document the plan and sell it to the company executives.

Writing the Plan

In general, a pollution prevention plan summarizes the results of the opportunity assessment and provides a schedule for project implementation. The schedule is necessary for monitoring and maintaining the overall program. If the team does not have the authority to commit funds, the plan may also be needed to secure funding for projects that require capital investment.

The contents of the plan, for each project, should discuss the project's potential for preventing pollution, the maturity of the technology and a discussion of successful applications, the comprehensive project economics, the required resources and how they will be obtained, and the estimated time for equipment installation and start-up. The plan should also include performance measures to allow the project to be evaluated after it is implemented.

Final corporate decisions on which projects will be implemented and what schedule is appropriate are based on this written pollution prevention plan. Once a decision is made, everybody in the company is obliged to support the implementation and continuity of the plan.

Implementing and Monitoring the Plan

Plan development, implementation, and assessment of the achievements are all related activities, because plan monitoring provides a feedback for plan revision. Eventually, the revised plan will be used to further facilitate a more effective and efficient implementation. Planning is indeed a continuous process.

Higgins (1991)[149] reveals that successful implementation of a pollution prevention project relies on two factors, and failure occurs if one of these factors is missing. The first success factor is production personnel that are strongly motivated to implement and maintain the necessary changes. Project staff are motivated by management support of the project, recognition for successful efforts, operations personnel involvement in planning, leadership from a waste minimization believer, declining production costs, increasing productivity, allocation of waste disposal costs directly to production units, and educating the staff on the environmental effects of what they throw away.

The second success factor is the use of elegant but simple technologies. This means that the change is not drastic but evolutionary, equipment is adapted to local conditions, new equipment is reliable and easy to operate and maintain, and extensive training is provided for the operating staff.

Based on experience with successful and less successful pollution prevention projects, Higgins recommends that the company should use waste production data to focus pollution prevention efforts on sources of waste that are the most costly. Frequently, personnel do not realize the true cost of waste disposal. Just knowing the costs can be sufficient motivation for production staff to make changes that will reduce disposal costs.

The pollution prevention program does not end with implementation. Once a project has been implemented, it is equally important to monitor its progress. Project monitoring refers to measuring the progress against the project's goals. This monitoring process helps identify pollution prevention techniques that work well and those that do not. Monitoring requires the collection of relevant data on the project to determine if the project had an effect on waste generation or other emissions.

Quantitative evaluation as part of the monitoring program is necessary to compare a facility unit with similar units in the company and with data from other companies. This quantitative analysis can be used to examine potential pollution prevention options for more successful projects and to enhance the existing program.

The success of implementing pollution prevention projects is undoubtedly dependent on the degree of support of employees for the projects. Therefore, communicating the plan to employees and obtaining support from them is mandatory.

Communicating the Plan

Many pollution prevention projects require changes in operating procedures, purchasing methods, materials inventory control, employee training, and documentation. The projects may also require new equipment or modifications to the existing equipment. Because of this changing work environment, communicating the plan, training, and incentive programs are needed to get employees used to the new procedures and new machinery.

The goal of plan communication is to keep employees aware, supportive, and motivated. They will take their pollution prevention roles more seriously when their managers keep them informed and encourage them to suggest pollution prevention ideas. Communication can be elevated through training and incentive programs.

The objectives of the training and incentive programs are:

1. to familiarize employees with their pollution prevention responsibilities,
2. to encourage employees to participate in pollution prevention,
3. to recognize and appreciate employees for pollution prevention efforts,
4. to keep employees informed of relevant environmental issues, and
5. to widely publicize pollution prevention success stories.

To achieve these objectives, the company must have a number of programs such as pollution prevention awareness, employee awards, and public outreach. A pollution prevention awareness program is designed to raise each employee's awareness of the company's commitment in pollution prevention. Training and sharing information represents the essential components of the program.

An employee award program includes recognition of employees who suggest pollution prevention ideas that prove feasible and are slated for implementation. The recognition should be internally publicized. Cash or other incentives can be awarded to individuals for outstanding accomplishments.

A public outreach program helps to demonstrate that the company's commitment to pollution prevention is genuine. The outreach can take the form of employees speaking at meetings of community organizations and at schools to publicize the company's pollution prevention programs. Papers written for conferences and journals add to positive publicity.

Maintaining internal communication is vital in pollution prevention. Employees will work more effectively when they know what management expects of them. Keep soliciting and following up on employees' suggestions. Forums such as breakfasts or meetings can promote the exchange of information that will generate new ideas.

Motivated employees will contribute significantly to the success of pollution prevention projects. Successful projects are good news for the company as well as for the employees. In many states, reporting pollution prevention success stories to the designated state agency represents a mandatory requirement for the regulated industry. The report must also include how the company measured the success. Please refer to Table 7.1 for a list of states requiring the submittal of a pollution prevention plan. The plans received from all companies are collected and compiled by the state environmental agency, then are analyzed and used as the basis for formulating and updating a state pollution prevention program.

State Pollution Prevention Program

State pollution prevention programs, in general, consist of either a voluntary technical assistance program or a pollution prevention planning requirement, or both. The scope and objective of a state program depends primarily on (1) the pollution prevention laws that exist, and (2) the extent of funding that is readily and sufficiently available to implement the laws. Normally, if a state has a mandated pollution prevention program, the funding is also made available.

The goal of a state pollution prevention program is to promote the use of pollution prevention to eliminate or reduce the generation and toxicity of hazardous waste and the use of toxic substance in a state. This goal will be achieved through the enactment of mandatory pollution prevention plan requirements and/or by providing voluntary technical assistance. Regardless of the strategy, a state pollution prevention program is essentially an educational undertaking. This educational endeavor can be done in several identifiable formats, such as providing a technical assistance program and other innovative programs, including assigning RCRA inspectors to promote pollution prevention through their inspection activities, or incorporating a pollution prevention commitment by RCRA violators in the compliance settlement language.

Examples of activities provided by a technical assistance program are:

1. dissemination of information through the "hands-on" pollution prevention publications, and the creation of a clearinghouse and forum for information exchange, including newsletters and electronic bulletin boards;
2. seminars of general pollution prevention interests, and workshops on specific processes, waste streams, or industries; and
3. on-site technical assistance.

Ferrante (1993)[150] reveals that a few state programs provide financial assistance to industries interested in pollution prevention. Some of the programs include waste-end taxes, loan and grant programs, and tax bonuses as additional incentives to pollution prevention. Ferrante also observes that state pollution prevention programs generally lack the funding necessary to perform thorough activities. Therefore, through a targeting effort using Form R or BRS data, states can identify and select areas on which to focus their resources.

The target of a state pollution prevention effort can be specific industries or specific pollutants. The state will provide assistance to any size company as requested, but will focus on specific industries or pollutants. The target can also be based on the size of the company. This targeting strategy can be critical, because small businesses are not ready to implement pollution prevention unless they are already incurring all the costs of complying with basic environmental regulations (Rozell and Brower, 1993).[151] Rozell and Brower underline that most small to mid-size facilities desperately need compliance assistance.

Small businesses usually qualify for SQG status. Mid-size companies may be classified as SQG or LQG, depending on the amount of hazardous waste they generate on a monthly basis. In terms of program planning, the above findings imply that an effective pollution prevention program would focus on its target group.

Educating SQGs on pollution prevention normally includes aspects of regulatory education as well. This means that when conducting a seminar for an SQG, it will be more effective if the seminar is designed to also invite RCRA and UST inspectors, used-oil staff, and air quality and wastewater treatment personnel as speakers. These regulatory staff can inform the small businesses on various pitfalls in complying with environmental regulations and how to avoid noncompliance.

Large companies are normally staffed with in-house pollution prevention experts. Assistance that can be offered to an LQG includes regulatory assurance that pollution prevention projects they undertake are indeed in compliance with existing environmental regulations.

Although assistance to the regulated industry represents the core of a pollution prevention program, assistance programs can create potential planning and management issues. This is particularly clear in the provision of an on-site technical assistance program. The following section examines this issue in more detail.

On-Site Technical Assistance

On-site technical assistance involves the dissemination of technical information to a facility in a detailed and site-specific manner. In general, on-site technical assistance is sought mostly by businesses that are serious about implementing pollution prevention but lack the necessary in-house resources to identify pollution prevention opportunities. State programs can fill this void by giving advice and guidance to these companies. However, when a program is administered by a regulatory agency, there are several issues that may need to be resolved, specifically those pertaining to trust, liability, and confidentiality.

With respect to **trust**, the issue is, will a facility be comfortable allowing regulatory staff on-site in a non-regulatory capacity, and what repercussions exist for a facility if a violation or noncompliance condition is identified during the visit? Does receiving on-site technical assistance present a risk of being cited for violations? With respect to **liability**, the issue is, who will be responsible for losses or damages that occur when a facility implements an opportunity identified during an on-site technical assistance and it does not work? The **confidentiality** issue is, what kind of guarantee can a public agency give to a facility that any proprietary or confidential information obtained during an on-site technical assistance will not be released without authorization?

Although there are no simple answers to these questions, many state programs have demonstrated that they are not really a cause for great concern. Some of the steps that are necessary for a government agency to deal with these issues are to contact the state attorney general's office for advice and recommendations, and to develop an agency policy that clearly addresses the role of state personnel during an on-site assistance meeting — including liability protection.

Once the policy is established, it is necessary to determine what types of assistance will be provided to the regulated industry. This decision must be carefully evaluated, because the service provided by the state agency may be in competition with similar services provided by environmental consultants.

The scope of on-site technical assistance can be decided by considering the successive evaluation procedure in the facility pollution prevention planning. As previously discussed, the procedure consists of three main steps:

1. the selection of potential targeted areas — either waste, process, or chemical, or any combination of them;
2. the identification of potential pollution prevention projects; and
3. the determination of feasible pollution prevention projects.

Depending on local conditions, an on-site technical assistance program can be restricted to step 1. The program may also cover steps 1 and 2, and leave step 3 open for the facility to pursue it internally or through the service of an environmental consultant. Most state technical assistance programs involve steps 1 and 2.

Some states utilize non-regulatory institutions, such as universities, to perform on-site technical assistance programs. This option is clearly advantageous for situations in which the element of trust represents a major barrier to state agencies providing on-site assistance directly to the industry. The drawback of the program is that the agency would not be in a better position to really appreciate the "in the field" pollution prevention problems.

Understanding problems and finding solutions are essential for the success of any program. In the field of pollution prevention, measuring the success of the program is generally not an easy task. Some of these difficulties will be addressed in the next section.

Measuring Success

Pollution prevention as a program can be viewed in terms of its characteristics as well as its ultimate goals. This distinction is significant when hazardous waste planners have to evaluate the accomplishments of the program. The premise of this book is that the main character of a pollution prevention program lies in its educational aspect. The purpose of educating the regulated industry is to raise awareness and increase motivation to implement pollution prevention.

Because it is difficult — if not impossible — to measure a facility's awareness or motivation, the normal yardstick to evaluate a program is the interest expressed in the state program. Indicators for such measurement can be obtained through opinion surveys. They can also be seen from statistics on the number of seminars given, participants, on-site technical assistance requests, requests for speakers, and other inquiries.

A much better indicator of whether a state is accomplishing its goal in preventing pollution, however, is the degree of actual reduction in the generation of wastes and toxic chemical usage. States normally use the biennial reporting system (BRS) and toxic release inventory (TRI) data base to evaluate accomplishments in these areas. Some states listed in Table 7.1 may utilize their own pollution prevention plan reporting system to generate information in addition to using the BRS and TRI reports.

The BRS data system provides pollution prevention information in Section IV of the generation and management (GM) form. The section requests that each company list any pollution prevention techniques that resulted in minimization of each reported waste. The section also requests a "yes" or "no" answer, which shows whether these pollution prevention techniques resulted in increased toxicity of the waste, or increased toxicity and amount of emissions into the air, water, or land.

In addition, the GM form provides, for each waste reported by a facility, the quantity of waste recycled and the quantity reduced during the reporting year. The quantity reduced is adjusted to incorporate changes in economic and other factors that affected the quantity of hazardous waste generated in the reporting year, and it is compared to the prior year. The adjustment is calculated by using the activity-production (A/P) index. Here is an example of how to use the index.

A firm manufactures tools using a process that generates a hazardous waste. In the reporting year, the firm produced 1200 tools and generated 1800 gal of waste. In the prior year, the firm produced 1000 tools and generated 2000 gal of waste. The actual waste generation change was from 2000 gal to 1800 gal — an actual reduction of 200 gal.

The A/P index is calculated as 1200/1000, or 1.2. This means that without pollution prevention activities, the quantity of hazardous waste that would have been generated is equal to 1.2 times 2000 gal, or 2400 gal. The adjusted waste generation change was therefore from 2400 gal to 1800 gal, resulting in an adjusted reduction of 600 gal.

The actual change measurement is straightforward, easy to understand, and relatively easy to calculate. The drawback is that the measure does not provide any insight into the causes of such changes in quantity (Baker, Dunford, and Warren, 1991).[152] The A/P index evidently provides a more accurate assessment of the change; however, the adjusted quantity change does not control for changes in operating conditions that may also affect the quantity of hazardous waste generated.

Baker et al. give the following example. If wastewater was generated at a constant rate, as long as the plant was operational, the index should have reflected changes in the hours the plant operated rather than changes in the level of product production. When completing the BRS, facilities often report the change in product production.

Even with this limitation, the BRS still represents the most widely used pollution prevention data source. The other data system that provides comparable information is the TRI report. The TRI form, also known as Form R, is a mandatory reporting format developed under Section 313 of the Emergency Planning and Community Right-to-know Act (EPCRA).

The form contains data on the quantities of TRI chemicals used and released into the various environmental media and what efforts are being made to reduce or eliminate them. This information provides a base line to assess progress made to prevent TRI chemicals from entering wastes, as well as progress made in the management of wastes containing TRI chemicals.

The universe of TRI chemicals, i.e., toxic chemicals regulated under TRI or EPCRA requirements, is a little broader than the RCRA regulated chemical constituents. However, the universe of facilities required to submit TRI forms is a little narrower than RCRA regulated industry, because TRI facilities are defined as those businesses that are included in the Standard Industrial Classification (SIC) codes 20 through 39 (i.e., manufacturing firms), have 10 or more full-time employees, and use at least 10,000 lb of the TRI chemicals *or* manufacture, process, or import more than 25,000 lb of the TRI chemicals annually. The TRI facilities are not required to submit Form R to state environmental agencies, unless required by state laws, therefore not all state environmental agencies have TRI data systems in place.

Implementing an effective state program is demanding. Planners have to create a good working relationship with the regulated industry, establish credibility with and obtain trust from the businesses, and solve quantification problems in measuring the progress. Adding the typical lack of state resources to this long list of issues makes state pollution prevention planning a challenge. Realizing the difficulty that each state has, the EPA recently initiated an incentive program to promote pollution prevention programs in states.

Pollution Prevention Incentives for States

Section 6605 of the 1990 Pollution Prevention Act (P2A) authorizes the EPA to make matching grants to states to promote the use of source reduction techniques by businesses. Less than a year after the enactment of P2A, in January 1991 the EPA formally announced its **pollution prevention strategy** — which discusses several activities that form the basis for implementing the EPA's program, including:

- identifying and overcoming obstacles to prevention,
- expanding public participation,
- enhancing state capabilities in pollution prevention,
- providing outreach and training, and
- using the existing regulatory framework to provide incentives for prevention.

The 1991 strategy also outlines the EPA's approach to facilitating pollution prevention initiatives in several sectors of the economy. This includes the EPA's **33/50 program,** which is directed at the industrial sector. The 33/50 program targets the reduction of 17 high-risk chemicals listed from the 1988 TRI data. The chemicals are:

1. benzene,
2. cadmium and compounds,
3. carbon tetrachloride,
4. chloroform,
5. chromium and compounds,
6. cyanide and compounds,
7. lead and compounds,

8. mercury and compounds,
9. methyl ethyl ketone,
10. methyl isobutyl ketone,
11. methylene chloride,
12. nickel and compounds,
13. tetrachloroethylene,
14. toluene,
15. trichloroethylene (TCE),
16. 1,1,1-trichloroethane (TCA), and
17. xylene.

By establishing the 33/50 program, the EPA is seeking voluntary commitments from industry to reduce the targeted chemicals by at least 33 percent by 1992 and 50 percent by 1995.

The Pollution Prevention Incentives for States (PPIS) was established in 1990. The history of this grant program began in 1989 when the EPA initiated financial assistance awards to state, tribal, and regional organizations to promote pollution prevention. When first introduced, the program was called Source Reduction and Recycling Technical Assistance (SRRTA). In 1990, SRRTA was renamed PPIS to acknowledge the EPA's increased emphasis on multimedia pollution prevention.

In 1990, 26 agencies received PPIS grants for a total of $7 million. In 1991, a grant of $4.5 million was shared by 16 agencies; in 1992, another 16 agencies received $3 million; and in 1993, the total grant award was $4.5 million (U.S. EPA, 1993).[153]

In evaluating grant applications, the P2A directs the EPA to consider whether the proposed pollution prevention projects will:

1. Make technical assistance available to businesses seeking information about pollution prevention opportunities, including funding for experts to provide on-site technical advice and to assist in the development of pollution prevention plans.
2. Target assistance to businesses for whom lack of information is an impediment to pollution prevention.
3. Provide training in pollution prevention techniques.

Eligible PPIS grant applicants are states and territories, including state universities and all federally recognized Indian tribes. Interested applicants should contact their EPA Regional Pollution Prevention Coordinator for more information.

CAPACITY ASSURANCE PLAN

At the state level, waste minimization and pollution prevention programs represent significant components of a Capacity Assurance Plan (CAP). A CAP is a planning document developed by a state to satisfy a federal requirement. The requirement is Section 104(c)(9) of the Comprehensive Environmental Response, Compensation, and Liability Act (CERCLA), as amended, which mandates states to assure the EPA that adequate capacity will exist to effectively manage all hazardous wastes that are reasonably expected to be generated within the state during the next 20 years.

Each state must develop a CAP, because no federal Superfund remedial action monies will be provided to a state unless it has provided such an assurance and the EPA deems the assurance adequate. Although the statutory provisions of the CAP is CERCLA, and the sanction of not developing a CAP is tied to a Superfund grant, the CAP is basically a hazardous waste planning document.

This section describes the methodology in developing a CAP document. In particular, the section discusses the concept of assuring capacity, what constitutes the CAP components, and some critical planning and policy issues in developing a CAP.

Assuring Capacity

Chronologically, the first submission of the CAP from states to the EPA was completed in 1989, the second CAP submission was in 1992, and the third CAP was in 1993. Although the CAP process and its scope of analysis have been modified during the past three submissions, the overall CAP principle is basically the same; i.e., the assurance status is evaluated by comparing the amount of waste generated and the availability and capacity of hazardous waste treatment and disposal facilities to manage the waste in the state.

The comparison will demonstrate whether or not a state has enough capacity to manage its own waste. If a shortage exists, the state is required to describe what efforts are to be made to address the deficiency.

Most CAP analysis is based on data contained in the biennial reporting system (BRS) data base.[154] Statutorily, the EPA requires the states to assure capacity for the next 20 years. For the 1993 CAP, the base year is 1991, because it is the most recent year for which states have collected BRS data. The 20-year projections start from the base year.

States can utilize data from the BRS and related software to help prepare the CAP's base year and projection tables (U.S. EPA, 1992).[155] For the base year and projection tables, states should provide capacity-related information for the 14 designated **CAP Management Categories**. The categories are:

1. Metals Recovery,
2. Inorganics Recovery,
3. Organics Recovery,
4. Energy Recovery — Liquids,
5. Energy Recovery — Sludges/Solids,
6. Stabilization/Chemical Fixation,
7. Incineration — Liquids/Gases,
8. Incineration — Sludges/Solids,

9. Fuel Blending,
10. Hazardous Wastewaters and Sludges Treatment,
11. Landfill,
12. Deepwell/Underground Injection,
13. Land Treatment/Farming, and
14. Transfer/Storage.

The above categories are defined in terms of the **Type Codes** as reported on the Generation and Management (GM), Waste Received (WR), and Process System (PS) forms of the BRS. The EPA developed the categories based on the following criteria (U.S. EPA, 1993):[156]

1. Each CAP Management Category consists of a number of waste management technologies that are generally interchangeable for managing *broad* types of waste (e.g., organics) based on treatment performance.
2. CAP Management Categories are grouped by recovery, treatment, and disposal technologies to reflect a preference for the national environmental management hierarchy.
3. Technologies for treating wastewater and the sludge resulting from wastewater treatment generally are similar (i.e., treatment tanks) and therefore are consolidated into one CAP Management Category.
4. The Transfer/Storage CAP Management Category was created because of the difficulties in determining the ultimate disposal of wastes exported to transfer facilities. This category is applicable only for exported waste presented in the base-year tables.

The relationship between the CAP Management Categories and the BRS Type Codes is provided in Appendix VII.2. The appendix shows, for example, that Type Codes M031 (acid regeneration) and M039 (other recovery — type unknown) are grouped in the Inorganics Recovery category; while the Metals Recovery category consists of Type Codes M011 (high temperature metals recovery), M012 (retorting), M013 (secondary smelting), M014 (other metal recovery for reuse), and M019 (metals recovery — type unknown).

The BRS Type Codes M135 (direct discharge to sewer under POTW) and M136 (direct discharge to surface water under NPDES) are not assigned to a CAP Management Category, because the systems manage wastes that are not defined as solid waste under 40 CFR 261.4(a). These two Type Codes are considered outside the scope of the CAP. If a BRS Type Code (such as M049 — incineration, type unknown) is applicable to more than one CAP Management Category, it is defined under all relevant categories (for M049, the relevant categories are Incineration — Liquids/Gases and Incineration — Sludges/Solids). When this circumstance occurs, states should use other BRS data, such as Form Codes, to determine the appropriate CAP Management Category.

The CAP base line tables provide information, by CAP Management Category, on the base line year's quantity of waste generation that will eventually go to commercial treatment and disposal facilities. This base line year amount represents the extent of "demand" for treatment or disposal capacities. Also provided in the tables is information on the annual capacity of commercial treatment facilities and the available capacity of commercial landfills in the state.[157] The capacity represents the available "supply" of capacity to manage wastes. The difference between demand and supply for each CAP Management Category will determine the adequacy status of each hazardous waste management capacity in the state. Because the adequacy of capacity is based on the capacity status of each CAP Management Category, this may create a circumstance in which a state may have sufficient capacity for landfill disposal but a shortfall in incineration or other treatment.

The adequacy status is assessed from the base year and the projection years. For the 1993 CAP, the projection years were 1999 and 2013. If a state does not have sufficient capacity in *any* of the categories in both the base year and the projection years, the state must address the shortfalls appropriately.

There are three alternatives that a state can use to address capacity shortage. The first two options are related to the demand and supply relationship. The logical way to balance the relationship is by reducing the demand side and/or increasing the supply side.

Reducing the demand for a CAP Management Category as the first option means minimizing the generation of hazardous waste. This option was discussed in the *waste minimization and pollution prevention* program in the preceding section. The second option is to increase the supply of a CAP Management Category by building new treatment and disposal capacities. This means increasing the capacity of the existing facilities and/or the siting of new treatment and disposal facilities. *Facility siting* is discussed in the next section. The third option is to arrange an agreement with other state(s) concerning the development of new capacity or waste minimization efforts through **interstate agreements**. The states may utilize one or any combination of the three options when addressing the shortfalls.

The procedure to address shortfalls in the 1993 CAP differs significantly from the procedure in the 1989 CAP. The 1993 CAP utilized a multiphase process by which a state with a deficiency in one CAP Management Category only had to address its deficiency *if* there was a national deficiency for that category. The 1989 CAP examined the capacity status of each state, and each state addressed any deficiency — regardless of the status of national capacity — by describing to the EPA the efforts to eliminate the deficiency. Some states grouped together and assured the capacity in a regional setting.

The CAP format is standardized by the EPA in order to facilitate uniform and consistent national evaluation of the CAP documents submitted by all states. The following paragraphs describe each of the CAP components that together make up the 1993 CAP document.

CAP Components

As a component of its CAP, a state must demonstrate an understanding of its hazardous waste generation and management system by providing information on the quantity of hazardous waste generated and managed in-state, and the amounts exported and imported. The CAP base year tables provide a framework for presenting this information. The base year is the year for which hazardous waste generation and management data are used to establish a base line for projecting future hazardous waste generation and management. The base year tables are generated from the BRS data base using the EPA's special software called "Table Talk."

The 1993 CAP requires states to prepare four base year tables presented below:

1. the hazardous waste generated and managed on-site,
2. the management of hazardous waste in captive systems,
3. the management of hazardous waste in commercial systems, and
4. the maximum operational in-state commercial hazardous waste management capacity.

The discussion of these base year tables is provided in the following section.

The Base-Year Generation and Management

The base year hazardous waste generation and management tables cover hazardous waste as defined under Subtitle C of the Resource Conservation and Recovery Act (RCRA); they also cover non-RCRA Subtitle C hazardous waste that is considered hazardous under respective state regulations and is managed in hazardous waste management systems. The quantity of hazardous waste generated and managed on-site is tabulated in a format specified in Table 7.2.

States are not required to demonstrate adequate capacity for hazardous wastes that are managed in on-site systems, because the EPA believes that the capacity needed to manage hazardous wastes on-site will continue to be available in future years. If a state has knowledge of a specific event that will cause a significant shift from on-site treatment and disposal to commercial management, this event should be considered in the commercial management projections. An example of this is when a state intends to close a currently operating landfill.

An on-site management system is a system that is only used to treat, dispose, or recycle hazardous waste that is generated on-site. If a generator has treatment, disposal, or recycling capacity available for hazardous waste received from other generators under the *same company ownership,* this generator basically operates a captive management system. The demand placed on a captive management system in the CAP base year table is divided into three groups (Table 7.3):

1. waste exported to out-of-state captive systems,
2. waste generated and managed within the state in captive systems, and
3. waste imported for management in captive systems.

As with on-site management, states will not be required to project wastes managed in captive systems. Additionally, if a state is aware of circumstances that

Table 7.2 1991 Hazardous Waste Generated and Managed On-Site (tons)

CAP Management Category	Waste Managed On Site
RECOVERY[a]	
Metals Recovery	
Inorganics Recovery	
Organics Recovery	
Energy Recovery - Liquids	
Energy Recovery - Sludges/Solids	
TREATMENT[a]	
Stabilization/Chemical Fixation	
Incineration - Liquids and Gases	
Incineration - Sludges/Solids	
Fuel Blending	
Hazardous Wastewaters and Sludges Treatment	
DISPOSAL	
Landfill	
Deepwell/Underground Injection	
Land Treatment/Farming	
TRANSFER/STORAGE	
Transfer/Storage	

[a] Data may not be complete for these technologies because facilities are not required to report in the 1991 Biennial Report waste managed in exempt processes.

From U.S. Environmental Protection Agency, *Guidance for Capacity Assurance Planning — Capacity Planning Pursuant to CERCLA §104(c)(9),* U.S. EPA, Washington, D.C., 1993.

would cause a significant shift from captive to commercial management, this shift should be considered in the commercial management projections.

A commercial hazardous waste system is defined as a system that has treatment, disposal, or recycling (TDR) capacity available to any hazardous waste generator. Included in this definition is limited commercial TDR capacity, which is available only to a limited number of generators. The demand placed on a commercial management system is divided into five groups (Table 7.4):

Table 7.3 1991 Management of Waste in Captive Systems (tons)

CAP Management Category	Exports	Waste Generated and Managed In State	Imports
RECOVERY			
Metals Recovery			
Inorganics Recovery			
Organics Recovery			
Energy Recovery - Liquids			
Energy Recovery - Sludges/Solids			
TREATMENT			
Stabilization/Chemical Fixation			
Incineration - Liquids and Gases			
Incineration - Sludges/Solids			
Fuel Blending			
Hazardous Wastewaters and Sludges Treatment			
DISPOSAL			
Landfill			
Deepwell/Underground Injection			
Land Treatment/Farming			
TRANSFER/STORAGE			
Transfer/Storage			

From U.S. Environmental Protection Agency, *Guidance for Capacity Assurance Planning — Capacity Planning Pursuant to CERCLA §104(c)(9),* U.S. EPA, Washington, D.C., 1993.

1. recurrent waste exported to out-of-state commercial systems,
2. one-time waste exported to out-of-state commercial systems,
3. recurrent waste generated and managed within the state in commercial systems,
4. one-time waste generated and managed within the state in commercial systems, and
5. waste imported for management in commercial systems.

Table 7.4 1991 Management of Hazardous Waste in Commercial Systems (tons)

CAP Management Category	Exports		Waste Generated and Managed In State		Imports[a]
	Recurrent	One-time	Recurrent	One-time	
RECOVERY					
Metals Recovery					
Inorganics Recovery					
Organics Recovery					
Energy Recovery - Liquids					
Energy Recovery - Sludges/Solids					
TREATMENT					
Stabilization/Chemical Fixation					
Incineration - Liquids and Gases					
Incineration - Sludges/Solids					
Fuel Blending					
Hazardous Wastewaters and Sludges Treatment					
DISPOSAL					
Landfill					
Deepwell/Underground Injection					
Land Treatment/ Farming					
TRANSFER/STORAGE					
Transfer/Storage					

[a] Imports cannot be divided into recurrent and one-time wastes due to limitations of information provided on Biennial Report WR forms.

From U.S. Environmental Protection Agency, *Guidance for Capacity Assurance Planning — Capacity Planning Pursuant to CERCLA §104(c)(9)*, U.S. EPA, Washington, D.C., 1993.

When states develop Table 7.4, they need to distinguish between recurrent versus one-time waste. The distinction is necessary because states are required to make projections on recurrent waste only, while the EPA will make projections for one-time waste. **Recurrent waste** is defined as waste generated from continuous and intermittent (e.g., leak collection and oil changes) processes, such as industrial processes. The CAP defines **one-time waste** as any contaminated materials or treatment residuals (soils, sludges, debris, equipment) generated by any of the following:

1. Superfund remedial actions,
2. state remedial actions,
3. Superfund removal actions,
4. corrective actions at RCRA hazardous waste management units,
5. closures of RCRA hazardous waste management units, or
6. other remediation activities, including those resulting from state and private emergency removals, environmental audits, and property transfers.

Planning activities related to one-time waste, i.e., a remedial action plan, is discussed in Chapter 8. States are not required to identify between one-time waste and recurrent waste for imports, because they do not need to assure capacity for imported waste. Also, imported hazardous waste cannot be separated into recurrent and one-time waste using the 1991 BRS information.

The summary of the maximum operational in-state hazardous waste commercial management capacity is tabulated in a format displayed in Table 7.5. Information for this table can be retrieved from the Process System (PS) form of the BRS. If certain data elements on the PS form are not available, states should use their hazardous waste permit data.

The definition of the maximum capacity of a hazardous waste treatment, disposal, or recycling unit depends on whether the unit is a landfill or a flow system (treatment and recycling). The maximum capacity of a landfill is defined as the quantity of waste that could enter the landfill over its remaining lifetime. The CAP defines the maximum capacity of a flow system as the maximum quantity of waste that could have entered the system in one year, assuming that the following conditions apply:

- no change in equipment
- an unlimited supply of waste similar to that managed in the base year
- willingness to add possible additional shifts
- routine down-time
- no impact from other systems that share the same unit
- permit and regulatory limits are not exceeded

Tables 7.2 through 7.5 illustrate the status of a state's hazardous waste generation and management in the base year, either on-site, by captive management systems, or by commercial hazardous waste facilities. Analysis of these tables will show the capacity status in the base year of a state, i.e., whether the state has excess capacity or a shortfall in any of the 14 CAP Management Categories. The EPA also requires that states provide the projected capacity status for the next 20 years. The demand and capacity projection is only for commercial recurrent hazardous waste.

Table 7.5 Maximum Operational In-State Commercial Subtitle C Management Capacity — End of 1991 (tons)

CAP Management Category	Maximum Operational In-state Commercial Subtitle C Management Capacity
RECOVERY	
Metals Recovery	
Inorganics Recovery	
Organics Recovery	
Energy Recovery - Liquids	
Energy Recovery - Sludges/Solids	
TREATMENT	
Stabilization/Chemical Fixation	
Incineration - Liquids and Gases	
Incineration - Sludges/Solids	
Fuel Blending	
Hazardous Wastewaters and Sludges Treatment	
DISPOSAL	
Landfill	
Deepwell/Underground Injection	
Land Treatment/Farming	
TRANSFER/STORAGE	
Transfer/Storage	

From U.S. Environmental Protection Agency, *Guidance for Capacity Assurance Planning — Capacity Planning Pursuant to CERCLA §104(c)(9),* U.S. EPA, Washington, D.C., 1993.

The Projections of Commercial Management Capacity

As indicated previously, the base year for the 1993 CAP is 1991. The projection years selected are 1993, 1999, and 2013 (Table 7.6). The EPA requires that the projections include the impact of EPA and state regulations that are finalized before the end of 1992, but exclude the effects of economic changes. The EPA made this decision based on preliminary analyses, which have shown that the effects of economic changes on waste generation will be subsumed by the effects

Table 7.6 Demand for Commercial Hazardous Waste Management Capacity from Recurrent Waste Expected to be Generated In-State (tons)

CAP Management Category	Baseline	Demand for Commercial Subtitle C Management Capacity		
		1993	1999	2013
RECOVERY				
Metals Recovery				
Inorganics Recovery				
Organics Recovery				
Energy Recovery - Liquids				
Energy Recovery - Sludges/Solids				
TREATMENT				
Stabilization/Chemical Fixation				
Incineration - Liquids and Gases				
Incineration - Sludges/Solids				
Fuel Blending				
Hazardous Wastewaters and Sludges Treatment				
DISPOSAL				
Landfill				
Deepwell/Underground Injection				
Land Treatment/Farming				
TRANSFER/STORAGE				
Transfer/Storage				

From U.S. Environmental Protection Agency, *Guidance for Capacity Assurance Planning — Capacity Planning Pursuant to CERCLA §104(c)(9),* U.S. EPA, Washington, D.C., 1993.

of new hazardous waste regulations and waste minimization and pollution prevention programs. Eventually, the question of whether a relationship exists between economic change and hazardous waste generation will need to be answered.

The 1993 projection method for regulatory change is determined as follows:

1. Identify the EPA waste codes that will be affected by regulatory change. For example, in the 1993 CAP, states accounted for impacts of certain land disposal restrictions (LDR).
2. Determine the quantity of these wastes generated in the base year.

3. Identify how these wastes and their residuals will be managed in the projection years, and estimate their demand on commercial capacity.

To facilitate the projection of recurrent wastes, the EPA has provided a list of the EPA waste codes that are affected by regulatory changes for the 1993 estimation. The EPA also provides a calculation procedure for states to estimate the demand on commercial capacity (U.S. EPA, 1993).[158]

For the 1999 projection, the EPA assumed that demand would be constant from 1993 to 1999. For the 2013 projection, the EPA decided that demand would be consistent from 1999 to 2013. The decision was made because the EPA believes that a detailed 20-year projection would be too inaccurate to provide a reasonable picture of future hazardous waste management.

Similar to the demand table, the projection years for estimating the supply of commercial management capacity are 1993, 1999, and 2013 (Table 7.7). Activities that will be reflected in the 1993 projection include adjustments for capacity that has become operational or closed, and a depletion of landfill capacity since 1991.

For the 1999 projection, the capacity supply is held constant from 1993 to 1999, except for landfill or if the state knows of changes that will increase or reduce capacity — such as building or closing a new facility. The landfill capacity is declining over time. The EPA provides an equation to calculate this landfill capacity change. The equation is:

$$\text{Capacity Change} = 6 \times [(1993 \text{ demand} + 1999 \text{ demand})/2]$$

The equation assumes that the average demand over the six-year period is the average of the 1993 demand and the 1999 demand.

For the 2013 projection, the supply of treatment capacity is held constant. For disposal capacity, i.e., landfills, the supply calculation can be done by using the capacity change equation appropriately.

Once all six tables are completed, they should be collated into a single document with a submittal letter signed by the authorized state official and transmitted to the EPA by the due date. The EPA decided that for the 1993 CAP, the document submission is divided into three phases. The overall process described above represents the Phase 1 CAP submittal. The Phase 2 CAP focuses on the national capacity status determination. The Phase 3 CAP addresses how a state will address its portion of a national shortfall in any CAP Management Category.

Determining Capacity Status

After states transmit their Phase 1 CAP submittals, the EPA will aggregate demand and capacity at the national level to determine, for each CAP Management Category, if sufficient national hazardous waste management capacity exists for the 20-year projection period. The EPA's demand projection is based on the following equation:

Table 7.7 Expected Maximum In-State Commercial Subtitle C Management Capacity (tons)

		Maximum In-state Commercial Subtitle C Management Capacity		
CAP Management Category	**Baseline**	**1993**	**1999**	**2013**
RECOVERY				
Metals Recovery				
Inorganics Recovery				
Organics Recovery				
Energy Recovery - Liquids				
Energy Recovery - Sludges/Solids				
TREATMENT				
Stabilization/Chemical Fixation				
Incineration - Liquids and Gases				
Incineration - Sludges/Solids				
Fuel Blending				
Hazardous Wastewaters and Sludges Treatment				
DISPOSAL				
Landfill				
Deepwell/Underground Injection				
Land Treatment/Farming				
TRANSFER/STORAGE				
Transfer/Storage				

From U.S. Environmental Protection Agency, *Guidance for Capacity Assurance Planning — Capacity Planning Pursuant to CERCLA §104(c)(9),* U.S. EPA, Washington, D.C., 1993.

Projected National Demand = projected recurrent demand, minus 10% for waste minimization, plus projected one-time demand

While states are required to make projections of recurrent demand, the EPA is responsible for projecting one-time demand. The one-time demand projection is a projection of one-time waste generation, which is based on analysis of five data sources: Superfund remedial actions, Superfund removal actions, RCRA corrective actions, underground storage tank (UST) cleanups, and state and private cleanups (U.S. EPA, 1994).[159]

The EPA identifies national capacity status by comparing projected national demand to the projected national capacity for each CAP Management Category. If adequate capacity exists nationwide for all CAP Management Categories, the EPA will not require Phase 2 or 3 documentation from any state. If shortfalls exist for any of the categories, the EPA will identify the states that contribute to the shortfall and require them to address the shortfall in Phase 2 of the CAP process.

Addressing Shortfalls

States with shortfall capacity must develop a Phase 2 CAP, which includes waste minimization and in-state commercial capacity information for each shortfall category. The waste minimization information must be provided by states claiming more than a 10 percent waste minimization projection, including any interstate agreements for collective waste minimization planning. The in-state commercial capacity information identifies capacity that is permitted but not yet operational or that has been issued a draft permit.

In Phase 2, states are not required to provide a specific amount of the shortfalls; rather, they should provide the EPA with information on new capacity and their best estimate of waste minimization efforts. The Phase 2 submission is used to adjust the previous national aggregation. If adequate capacity exists in all CAP Management Categories after the adjustment, Phase 3 will not be required. If national shortfalls are still projected for any category, the EPA will identify those states that contribute to the shortfalls. These states are referred to as "shortfall states."

Shortfall states will be assigned a portion of the shortfall based on their demand for the CAP Management Category. Each shortfall state must deliver a Phase 3 CAP that outlines the steps the state will take to assure capacity for its portion of the national shortfall through increased waste minimization, interstate agreements concerning increased waste minimization, development of new capacity, and interstate agreements concerning the development of new capacity.

If a state is identified as having to address capacity shortfalls, the state should include milestones for each shortfall CAP Management Category in its Phase 2 or 3 CAP submittals. A milestone is a task or achievement that is scheduled to be accomplished by a specific time to ensure adequate capacity. A state plan to enhance waste minimization targeted at a specific waste stream, for example, should have milestones for activities needed to accomplish the program objectives. If a shortfall state is planning to develop a new waste management capacity, it should have milestones for the issuance of draft permits. The EPA tracks the milestones through the end of the projection year.

As we can see from the preceding paragraphs, the development of CAP documents is complex. Capacity assurance planning is essentially a continuous planning effort. States are expected to submit new CAPs on a regular basis to remain eligible for Superfund remedial monies. During the planning process, the states have the opportunity to identify issues and develop recommendations to the EPA to improve the planning process for the next CAP cycle.

Planning Issues in Developing the CAP

Public confidence in the CAP depends upon the credibility of the CAP methodology and the validity and reliability of the CAP data. When evaluating the 1989 CAP documents, the U.S. General Accounting Office (U.S. GAO, 1991)[160] identified serious problems in the methodology and data collection need in developing the state CAPs. The GAO provided a list of recommendations to address some of the problems.

In response to concerns about the 1989 CAP, the EPA developed and improved the CAP methodology by appropriately incorporating the recommendations from the GAO, the National Governors' Association (NGA), various organizations, and the public. The result is that the 1993 CAP is methodologically better than the 1989 CAP.

Issues related to the CAP data, however, are still not resolved completely because of the existing regulatory limitations and the variation in the level of stringency in state hazardous waste management programs. In this section, selected data management issues will be briefly discussed. The discussion will focus on the following issues:

1. double counting and underestimation of capacity demand,
2. import-export data discrepancy and reconciliation, and
3. waste minimization program assessment.

In order to understand the double counting issue, we must be aware that, as part of the course of transportation, transfer facilities receive hazardous wastes from generators and then ship these wastes to treatment or disposal facilities. Double counting occurs when the amount shipped by a transfer facility is counted under waste generation by both the transfer facility and the generator that originally produced the wastes. The EPA requires the states to disregard waste quantities shipped by transfer facilities to *avoid double counting* when completing the CAP.

Hazardous waste included in the CAP consists of: (1) RCRA Subtitle C hazardous waste, and (2) non-RCRA Subtitle C waste that is considered hazardous under state regulations (known as "state only" waste) and is managed in hazardous waste management systems (e.g., hazardous waste landfills). EPA requirements specifically exclude hazardous waste generated by small-quantity generators (SQGs).

Other types of generators that are not considered are conditionally exempt small-quantity generators (CESQGs) and households that produce what is known as household hazardous waste (e.g., paints). CESQG wastes are normally accumulated in a hazardous waste transfer facility for transport to a TSD facility. Household hazardous wastes (HHW) become municipal solid waste, unless local governments (cities or counties) provide a household hazardous waste collection center or a collection program. Once collected, the wastes are shipped to a transfer facility and eventually to a hazardous waste TSD facility (although legally this waste is still defined as solid waste).

Clearly, a transfer facility represents an important data source regarding wastes generated by SQGs and CESQGs and possibly HHWs. By disregarding waste quantities shipped by transfer facilities to treatment or disposal facilities, however, the CAP disregards the potential use of available hazardous waste management capacity by these small generators. The exclusion results in an *underestimation of the demand* for capacity. The magnitude of waste generated by these small generators is relatively unknown, but the quantity can be significant. In Chapter 6, the amount of hazardous waste generated by SQGs in Arizona in 1987 was estimated to be $^1/_8$ of the amount generated by LQGs. An estimate of HHW generated in that state in 1990 shows that the amount of HHW could be in the range of 20 to 40 percent of the amount generated by LQGs (Soesilo, 1992).[161]

Clearly, the development of accurate data for the CAP is dependent upon avoiding double counting and not underestimating the capacity demand. Unfortunately, the accuracy of the data is limited by regulatory constraints (i.e., SQGs, CESQGs, and HHWs are not required to comply with the BRS requirements) and by the difficulty in obtaining data from these small generators.

Import-export data represent another planning issue. We showed in Chapter 6 that because of the inclusion of state-only waste in the definition of hazardous waste in a state, the state may have discrepancies with other states in the import-export data. Also, the data base used to generate import-export data can be different (e.g., the use of BRS versus the use of manifest). These two factors alone may create *difficulties in reconciling* one state's import-export data with other related states'. It is predicted that this type of difficulty will not easily be solved.

Finally, there is the issue of waste minimization assessment. The EPA evaluates the success of a state waste minimization program through four criteria, which together work as an integrated system — each criterion carrying equal weight. The EPA uses these criteria as a standard of fairness and accuracy in evaluating the CAP. The criteria are:

1. adequacy of the state waste minimization infrastructure,
2. targeting of waste reduction activities,
3. generator communication, and
4. feasibility.

The EPA assumes that a 10 percent waste minimization is attainable nationally; therefore, states are given the 10 percent waste minimization credit in their CAPs. States are required to provide limited documentation only if they claim 10 to 20 percent waste minimization. If the claim is between 20 to 50 percent, the states are required to respond more fully. More detailed documentation is required if the claim exceeds a 50 percent waste reduction from minimization.

In order for the EPA to evaluate the *adequacy of the state waste minimization infrastructure,* a state should address, among other things, the types of activities included in the state waste minimization program, including its statutory authority; its budget appropriation, staffing level, and allocation of resources into program activities; its statewide waste minimization goals; its existing and planned system; and its method to monitor the program. With respect to the second criterion, the measure that the EPA utilizes to verify that the states are reasonable in **targeting waste minimization** is whether a state has a mechanism to show that the targeting is based on documented data sources and an outreach and technical assistance program to promote waste minimization. The EPA provides the third criterion based on its belief that the estimated future levels of waste reduction will be more credible if generators *participate in and commit to* the waste minimization projection. The states should provide documentation of the commitments. The feasibility measure is imposed to see if the waste minimization *projections are realistic,* technologically and economically attractive, and attainable.

Clearly, the review mechanism of the program enables the EPA to evaluate the level of state commitment to its waste minimization program. The mechanism is a good tool to evaluate the success of the state program. A successful state program may not necessarily result in successful actual reduction of the waste. This actual reduction can be assessed through monitoring the state's waste minimization milestones, which is required by the EPA as part of the CAP submittals. In reality, the actual minimization of the wastes in a state is an aggregate of minimization results of all contributing generators; therefore, the problem in quantifying the success of the actual minimization discussed in Chapter 6 is also applicable here at the state level.

We know from the preceding paragraphs that besides the waste minimization program, the other CAP alternative to address shortfalls of capacity is through the development of new hazardous waste capacity. This alternative can be fulfilled by expanding the capacity of the existing treatment and disposal facilities operated in a state or by building a new facility in the state. This topic is described next.

FACILITY SITING

Facility siting has always been a major issue in the field of hazardous waste management. The siting, construction, and operation of hazardous waste management facilities were historically a private-sector activity. Industries that generate hazardous wastes from their business operations might build storage, treatment, or disposal facilities on their property to accommodate their waste management needs. Alternatively, they might ship their hazardous wastes to commercial TSD facilities because of the cost-effectiveness calculations. The generators and the TSDs typically have made their own decisions regarding where to dispose of wastes and what waste management technologies to build. State governments did not get involved in those activities, except to oversee that the activities complied with the laws.

With the growing public concern about the potential risks associated with hazardous waste, the NIMBY syndrome, and — most recently — the CAP requirements, many state governments currently have an increased role in the siting process. Hazardous waste facility siting is essentially an interplay among the industry, the government, and the public. Kusterer (1991)[162] observes that a successful siting is always characterized by, among other things, involvement of elected officials and good communication with the citizens.

State Roles in the Siting Process

As expected, state roles in the siting process vary from state to state. Apogee Research Inc. (1988)[163] provides a variety of siting factors that can be used to explain variations in the siting practices among the states. The factors include:

1. The availability of the state legislation to address the siting.
2. The authority of a state to site a hazardous waste facility. A few states (e.g., Arizona, Colorado, and Minnesota) are mandated to select a site; however, some have a mandate to choose a site only if the private sector fails to site the facility.
3. The availability of a state hazardous waste management plan. This plan is normally required by state laws.
4. The availability of provisions regarding negotiation, mediation, or arbitration between the facility and the host community.
5. The availability of a state policy that requires host community representatives on the siting board. The representation can be an advisory role, or the policy may allow the representative to have the power to negotiate.
6. The availability of technical assistance or a state grant fund to help defray the cost of site review by the host community.
7. The mandate of a state to provide economic incentives to encourage local communities to accept a hazardous waste management facility.

The first factor, the **state siting law**, describes a state's approach to the siting of new hazardous waste management facilities. The scope and details of the statutes vary from state to state. For example:

- Some statutes (e.g., Kansas, Pennsylvania) may require the state to conduct studies related to the hazardous waste management needs of the state and to develop a statewide plan for managing hazardous wastes.
- Some statutes (e.g., Indiana, New Jersey, North Carolina) establish a formal siting board to review siting proposals and to institute a process for submitting, reviewing, and deciding on siting applications.
- Many statutes (e.g., Maryland, Nebraska, Michigan) mandate the public involvement procedure adopted in the siting process.
- Some statutes (e.g., Kentucky, Massachusetts) require some sort of compensations to the host communities.
- Many statutes (e.g., Utah, Florida) provide a list of technical criteria that address safety and risk to human health and the environment. These criteria for selecting a site include the proximity of the site to residential

areas, the possibility of a flood or an earthquake at the site, the probability of contaminants reaching groundwater, and the risk of transporting the wastes to and from the site.

Given the variations of siting practices among the states, Houghton (1989)[164] suggests that the siting program status of a state be evaluated based on the following three criteria:

1. Who initiates the siting proposals?
2. Who makes the final decision on whether a proposed facility should be approved?
3. How is public participation in the siting process structured?

The *initiation of siting proposals* can be seen from the two extremes. On one hand, states are waiting for the private sector to approach them for permission to operate. Under this scenario, the state's involvement may be to just issue a hazardous waste facility operating permit. On the other hand, states are acting as the major initiator, offering sites suitable for private commercial hazardous waste facilities, or even buying sites for state-owned facilities. Please keep in mind that only states with an explicit legislative mandate may initiate siting proposals.

Most states are found between these two extremes. These states are trying to facilitate and encourage private siting initiatives through a variety of activities. The activities may include developing an inventory of potential sites or providing information on what type of facilities are needed in the state.

The authority to *approve or disapprove a siting proposal* in most states (with siting legislation) is with the state. Some states, such as Colorado and Florida, have *override* powers, i.e., the authority to overturn or overrule a local community's decision to prohibit the construction of a hazardous waste management facility. Some states have *preemption* powers, i.e., the authority to preempt local zoning authority on its decision as to whether the facility should be built on the proposed site. While in the override process the state steps in after a local community has made a decision, in the preemption process the state prevents the local community from the decision-making process altogether. Some other states are authorized to use their powers of *eminent domain,* by which the state can acquire property on which to build a facility providing there is just compensation for the property.

Public participation, as the third criterion in evaluating a state siting program status, represents the most important element. There is no consensus among the states on what form such participation should take. The disagreement is, for example, whether or not the participation should be facilitated from the start by giving the public a strong voice in the planning to the final say over whether a facility is built. Others argue that there is no such thing as full public participation; at some point, the state must be able to step in and act in the interest of the state as a whole and, if necessary, go against local community wishes.

Houghton identifies five forms of public participation. They are:

1. Local representation on planning or siting boards. This is temporary representation dealing with capacity to hold a hazardous waste facility. The representatives participate only in decisions relevant to their particular community and serve on the board while their community is under consideration to host a hazardous waste management facility.
2. Public participation in the state's hazardous waste management planning. This is full representation. The representatives participate in the formulation of the state hazardous waste management plan and represent the general public of the state, not simply the parochial interests of their local community.
3. Technical assistance to communities. An example of this effort is a grant award from the state to a host community to hire an independent consultant to review the technical aspects of the proposed facility, to help local residents understand the complex technical features, and to enable them to make an informed assessment of the risks. The residents are more likely to trust the review of an independent consultant than that of the state or industry.
4. Negotiation, i.e., the process whereby supporters and opponents of a facility sit together to resolve their differences. The objective of negotiation is to arrive at a remedy that prevents or corrects real or perceived problems arising with the construction and operation of a new facility.
5. Compensation, i.e., to provide "economic incentives" to a host community, independent of any agreements reached during negotiations. An example of compensation is when a state provides a direct appropriation to a host community accepting a hazardous waste management site. Some states have provisions allowing host communities to tax the facility's gross receipts. Non-monetary, material compensation, such as the purchase of a new fire truck for the host community or assurance to guarantee the availability of potable water, are common forms of compensation. The disadvantage of compensation programs is that they may be viewed as a form of bribery.

The success of facility siting depends on, among other things, the approach used in the siting process. After an analysis of siting attempts in the U.S. between 1980 and 1986, the National Governors' Association (NGA) concluded that there is no single siting approach that is workable in all states. Apogee Research Inc. (1988)[165] reveals, however, that there are some factors that seem to be associated with successful siting. Among others, the factors are as follows:

1. The rate of success for on-site facilities was greater than for commercial facilities.
2. Sites proposed on already industrialized land were more acceptable than those on nonindustrial areas.
3. Storage, transfer, and mobile treatment facilities were easier to site than other types of facilities.

The above findings suggest that a commercial hazardous waste management facility is the most difficult facility to site. In order to enhance the chances of siting commercial facilities, states should take steps to ensure that maximum use is made of public participation techniques.

Toward a Successful Siting

In September 1992, the NGA convened a group of state policymakers, representatives from the federal government, industry, academia, and environmental groups to discuss what is perceived as gridlock in the siting process. Becker (1993)[166] reports that there was no magic formula for successful siting presented by any state; however, seven guiding siting principles were most often mentioned during the conference. The principles are:

1. *Ensure that the host community will be brought into the process every step of the way.* Such involvement can promote confidence and trust, and the developer of the facility can meet the community's needs head-on instead of wasting time addressing opposition as the project develops.
2. *Establish evidence of need for the facility.* A host community might be more open to a hazardous waste management site if its members recognize that the proposed facility will meet the needs of local or regional waste generators, and if they are given assurance that their local needs are being met first.
3. *Address issues important and relevant to the community when discussing risk.* Communities want to know what impact the proposed facility will have on their standard of living. Risk assessment procedures must not ignore the community's perceived risk. Communities normally are concerned about the risks of transporting hazardous waste and materials, accidental spills or releases, decreased property values and business development, and reduced health and safety.
4. *Realize that communities are interested in empowerment.* The idea of providing compensation is good but probably not realistic. From the very beginning of the process, the community should be given a voice in how the project operates; therefore, its members should feel they have control over the process.
5. *Consider the cumulative effects of siting decisions.* This is especially important where some communities may bear a disproportionately larger environmental burden. In reality, these communities tend to be in low-income or minority areas. Concern about this environmental equity issue may force state and site developers to consider developing new sites in relatively "cleaner" geographic areas — a choice not likely to make siting easier.
6. *Generate support for waste management through public education.* The goal is to increase community awareness of the need for waste

management facilities as well as the risks and benefits involved. The waste generators can play a significant role in this effort, because they are the ones that make the connection between jobs and waste.

7. *Provide a clear state siting and permitting process.* The goal is to avoid an expensive and time-consuming process that can be a deterrent to many site developers.

Each of the principles discussed signifies the growing importance of public participation in a siting effort. With careful siting strategy, it is expected that more successful siting can be accomplished in the future. In case of a siting conflict, the best remedy seems to be through negotiation, because negotiations create an incentive for people to bargain and make tradeoffs until they are genuinely satisfied (Sanderson, 1985).[167] Negotiation remedies for siting conflicts can take many forms.

- It can be an action, or a promise of future actions.
- It can be the inclusion of additional safety features in the facility design, or restrictions on operating hours and procedures.
- It can be limitations on types and volumes of wastes handled at the facility.
- It can be the establishment of a program to monitor for health and safety violations.
- It can be property value guarantees or cash compensation to the community.

In relation to conflict remedy through negotiations, Bingham and Miller (1984)[168] provide a strategy to perform negotiations. They first argue that attempting to resolve hazardous waste facility siting disputes through negotiation remains a difficult proposition. Frequently, the issues that must be resolved first are the *need for the facility* and the *choice of technology.* These two issues can be difficult to negotiate particularly at the local level. For negotiations to be successful, there must be **incentives** to negotiate for all concerned. In reality, states that did not have an abundance of public participation forums, but *did* have negotiation or incentives programs, reported greater success in issuing permits to new facilities than those states with several types of public participation but no negotiation or incentive components (Daly and Vitaliano, 1989).[169]

Gregory and Kunreuther (1990)[170] argue that all sides to a negotiation must understand that refusing a proposed hazardous waste management facility incurs costs stemming from increasing waste disposal fees and the potential of illegal dumping. The consequences of this status quo option must be clearly stated. What counts is not just the risks to the population of a single community but the magnitude of the risks faced by the broader population at large. A community with a stake in the siting decision has to judge the facility by weighing the incentives offered if siting is accepted against the costs of the default option.

According to Bingham and Miller, developers who wish to attempt negotiation might consider making conditional offers to the host community as an incentive to negotiate. These might include promises that if an agreement were reached, it would include protection of property values, continuing communication with the community throughout the life of the facility, or something else that the developer knew was of concern to the community. Without specifying particular offers too soon, a developer could reassure the community that the facility was offering an incentive for negotiation.

Once the local community agrees to negotiate, the developer and the community must give adequate time and attention to preparing for negotiations. The variety of topics that must be addressed include determining who should be at the table, what issues are on the agenda, whether the meetings are to be open or closed, scheduling of the meetings, and other procedural matters.

The host communities are normally not monolithic; therefore, care should be taken in selecting a negotiating team to ensure that the residents feel that there is someone on the team who can represent their views. Preparations for negotiation should include developing internal, team bargaining, and consensus-building relationships prior to and during negotiations with the developer.

The importance of the negotiation process is that it requires the ability to identify what each side needs to achieve before it can accept an agreement. The process is not ended when an agreement is reached and ratified; attention must be paid to how the agreement will be implemented. Keep in mind that in hazardous waste facility siting, there will always be a continuing relationship between the facility operator and the local community.

Bingham and Miller caution that more experience with negotiated siting agreements is needed before conclusions can be drawn about whether negotiation approaches will work in practice and what procedures will be the most effective. This means that facility siting remains a challenge for planners now and in the future.

Facility siting, the Capacity Assurance Plan (CAP), waste minimization, and pollution prevention are all closely related activities in state hazardous waste planning. In the next section, we describe in detail what constitutes state hazardous waste management planning, its process, and its current status.

STATE HAZARDOUS WASTE MANAGEMENT PLANNING

When the CAP documents were developed and finalized by the states, there was an argument over whether or not the CAP could be called a state hazardous waste management plan. The consensus was that it was not. The CAP is more appropriately categorized as a planning tool. The states, however, have had mixed opinions about whether the CAP is an effective planning tool (U.S. EPA, 1993).[171] Most states agreed, however, that the CAP process provides useful forums for discussing waste management needs and plans. Many states thought that the states were developing better hazardous waste management plans through their own initiatives outside the CAP process.

In general, industry views the CAP process as both good and bad (Environmental Resources Management, 1989).[172] The CAP can be destructive insofar as it encourages states to erect barriers to interstate commerce and to balkanize the national markets. It can be constructive if it increases the state's understanding of the market and promotes state cooperation to overcome the NIMBY problem. The states are supportive of the CAP process because the CAP improves data management, promotes waste minimization, improves communication with industry, promotes industrial development, promotes interstate communication, and encourages regional dialogue.

Although the CAP is developed to anticipate a long-term solution to assuring hazardous waste management capacity, it represents only a fraction of a statewide hazardous waste management plan. A more comprehensive state hazardous waste management plan, or state plan, is needed. Currently, the state planning practice is still at its infancy stage. However, a recent trend suggests that the importance of state planning is becoming more apparent in the 1990s, which is particularly due to the increasing awareness of the public about environmental issues. The final part of this chapter is dedicated to the discussion of this type of planning.

State Planning Process

The purpose of state hazardous waste management planning is to develop strategic recommendations to all state participants in the hazardous waste management field. The recommended actions include plan implementation and plan monitoring, which are designed to protect human health and the environment and prevent unnecessary site remediation activities. The goal of state hazardous waste management planning is formulated by the state's legislature or is based on the state's needs.

Using the rational planning approach, the state hazardous waste management planning process can be viewed as consisting of the five following interactive stages:

1. identifying the most important needs, problems, and issues in the current state hazardous waste management practices;
2. gathering information to characterize the current situation and to project the anticipated future scenarios;
3. examining the plan's alternatives and specifying action plans to achieve the goals of state planning;
4. developing a planning report that describes the planning needs, analyses, findings, and recommended strategies for future actions; and
5. implementing the recommended actions and monitoring the results. Monitoring is needed because it is a tool for implementing changes and incorporating new ideas.

The above planning process begins with the identification of the hazardous waste management planning needs. The needs can be identified through a series of surveys, workshops, meetings, and discussions involving stakeholders in the hazardous waste management system — i.e., the hazardous waste handlers, the regulatory agencies, and the general public. Identifying planning needs, management problems, and societal issues may result in a broadening of the plan's vision and the refinement of planning goals.

The hazardous waste needs assessment represents a unique element of the planning process, because the formulation of the needs governs the scope and characteristics of the state hazardous waste plan. In other words, the more we know about the planning needs, the better we understand how to construct the state plan. This section illustrates the variety of hazardous waste management planning needs by summarizing the results of the EPA's study on this subject.

The study (U.S. EPA, 1993)[173] examines 18 selected states that are engaged in state hazardous waste management planning. The study shows that the needs that are identified by the surveyed states fall into six categories:

1. The need to manage hazardous waste according to the national environmental management hierarchy. This means that aggressively promoting waste minimization and pollution prevention programs represents the highest priority.
2. The need to understand the hazardous waste management capacity status of the state, i.e., to realize the demand for hazardous waste treatment and disposal services created by in-state hazardous waste generators and the availability of such services in-state.
3. The need to conduct public education in order to better inform the general public about hazardous waste risks and the means to address such risks. This includes informing the regulated community of opportunities to improve management.
4. The need to improve the implementation of the RCRA program, including better targeting of program efforts and increased staff training.
5. The need to promote compliance with hazardous waste regulations by increasing the rate of compliance with hazardous waste regulations on the part of the industry.
6. The need to balance environmental and economic considerations by incorporating specific economic considerations into regulatory and technical assistance activities.

The study reveals that, in general, the states placed primary emphasis on the first three categories. The first category, i.e., managing hazardous waste according to the hierarchy, reflects the importance of waste minimization and pollution prevention programs as seen by hazardous waste stakeholders. The second most important need is evaluating access to hazardous waste management capacity, which, for the most part, is triggered by the CAP requirements. Conducting public education represents the third most important need because of the apparent link between public education and achieving the objectives related to the other needs categories.

The identified planning needs are basically the same as the planning objectives that the state attempts to achieve. Once the needs or the objectives are identified, the next step is to describe planning practices designed to address those objectives.

State Planning Practices

The 1993 EPA analysis of state planning practices relied on a review of documents and discussions with representatives of the selected states. The study found that 28 separate planning activities were identified. The planning activities were conducted in order to achieve the six planning objectives. Three examples of planning activities are: (1) examining total waste generation by type and quantity for all wastes requiring RCRA hazardous waste management capacity, (2) assessing the life expectancy of hazardous waste treatment facilities, and (3) evaluating the most feasible management options for selected waste streams. In the following paragraphs, 15 planning activities to address the first three planning objectives are discussed.

Of the 15 planning activities, three are related to the first objective, i.e., **managing hazardous waste according to the hierarchy**. The activities are:

1. evaluation of the most feasible management options for hazardous waste streams. This is used to examine the potential for shifting from a lower priority in the hierarchy to a more preferred management category;
2. analysis of the economic and regulatory barriers to achieving the most preferred hazardous waste management category; and
3. identification of state procedures that promote hazardous waste management according to the hierarchy.

Nine other activities are related to the second objective, i.e., **assuring access to hazardous waste management capacity**. These activities are:

1. analysis of the quantity of waste generation by type for all hazardous waste requiring RCRA management capacity;
2. analysis of wastes of particular concern to the state, such as used oil or exempt wastewaters;
3. analysis of the spatial distribution of hazardous waste generators by type and size;
4. determination of the disposition of in-state hazardous waste by type and location of managing facility;
5. assessment of the life expectancy of in-state hazardous waste treatment, storage, disposal, and recycling facilities;
6. projection of future hazardous waste generation and management by considering waste minimization, pollution prevention, regulatory change, remediation wastes, and economic change;

7. projection of preferred future hazardous waste generation and management in accordance with the national hazardous waste management hierarchy;
8. determination of the balance between in-state hazardous waste management capacity and the demand for the capacity. This comparison is based on the preferred future projections; and
9. analysis of the economic viability of any capacity shortfalls as a result of activity #8.

In order to accomplish the third objective, i.e., **conducting public education**, the following planning activities are required:

1. surveys of generators and the general public to analyze their level of understanding of hazardous waste management systems, associated risks, and specific concerns that can be used to target educational efforts;
2. evaluation of opportunities to educate generators on waste minimization and pollution prevention; and
3. development of periodic publications for distribution to the general public.

The 1993 EPA study found that most of the state plans contain specific recommendations which many of the states have implemented. The recommendations cover a variety of implementation activities. The activities must be implemented by the state environmental agency, legislature, EPA, local governments, hazardous waste handlers, and the public (WSDOE, 1992).[174]

Although the implementation activities vary from state to state, the EPA study found that the activities can be categorized into three major groups: follow-up studies, legislative changes, and administrative changes. The follow-up studies are required as supporting documentation for monitoring and evaluating the plan. The state plan requires periodic evaluation and revision to remain as a dynamic and viable document.

Appendix VII.1 EPA Pollution Prevention Worksheets

Firm ____ Site ____ Date ____	**Pollution Prevention Assessment Worksheets** Proj. No. ____	Prepared By ____ Checked By ____ Sheet ___ of ___ Page ___ of ___

WORKSHEET 1

SITE DESCRIPTION

Firm: ____

Plant: ____

Department: ____

Area: ____

Street Address: ____

City: ____

State/Zip Code: ____

Telephone: ( ) ____

Major Products: ____

____

SIC Codes: ____

EPA Generator Number: ____

Major Unit: ____

Product or Service: ____

Operations: ____

____

____

____

____

____

Facilities/Equipment Age: ____

____

____

____

Appendix VII.1 EPA Pollution Prevention Worksheets (Continued)

Firm ____________ Site ____________ Date ____________	**Pollution Prevention Assessment Worksheets** Proj. No. ____________	Prepared By ____________ Checked By ____________ Sheet ___ of ___ Page ___ of ___

WORKSHEET **2**

PROCESS INFORMATION

Process Unit/Operation: ____________

Operation Type: ☐ Continuous ☐ Discrete ☐ Batch or Semi-Batch ☐ Other ____________

Document	Status					
	Complete? (Y/N)	Current? (Y/N)	Last Revision	Used in this Report (Y/N)	Document Number	Location
Process Flow Diagram						
Material/Energy Balance						
Design						
Operating						
Flow/Amount Measurements						
Stream						
Analyses/Assays						
Stream						
Process Description						
Operating Manuals						
Equipment List						
Equipment Specifications						
Piping and Instrument Diagrams						
Plot and Elevation Plan(s)						
Work Flow Diagrams						
Hazardous Waste Manifests						
Emission Inventories						
Annual/Biennial Reports						
Environmental Audit Reports						
Permit/Permit Applications						
Batch Sheet(s)						
Materials Application Diagrams						
Product Composition Sheets						
Material Safety Data Sheets						
Inventory Records						
Operator Logs						
Production Schedules						

Appendix VII.1 EPA Pollution Prevention Worksheets (Continued)

Firm ______	**Pollution Prevention Assessment Worksheets**	Prepared By ______
Site ______		Checked By ______
Date ______	Proj. No. ______	Sheet ___ of ___ Page ___ of ___

WORKSHEET
3

INPUT MATERIALS SUMMARY

Attribute	Description		
	Stream No. ____	Stream No. ____	Stream No. ____
Name/ID			
Source/Supplier			
Component/Attribute of Concern			
Annual Consumption Rate			
Overall			
Component(s) of Concern			
Purchase Price, $ per ______			
Overall Annual Cost			
Delivery Mode[1]			
Shipping Container Size & Type[2]			
Storage Mode[3]			
Transfer Mode[4]			
Empty Container Disposal Management[5]			
Shelf Life			
Supplier Would			
— accept expired material? (Y/N)			
— accept shipping containers? (Y/N)			
— revise expiration date? (Y/N)			
Acceptable Substitute(s), if any			
Alternate Supplier(s)			

Notes:
1. e.g., pipeline, tank car, 100 bbl tank truck, truck, etc.
2. e.g., 55 gal drum 100 lb paper bag, tank, etc.
3. e.g., outdoor, warehouse, underground, aboveground, etc.
4. e.g., pump, forklift, pneumatic transport, conveyor, etc.
5. e.g., crush and landfill, clean and recycle, return to supplier, etc.

Appendix VII.1 EPA Pollution Prevention Worksheets (Continued)

Firm ____ Site ____ Date ____	**Pollution Prevention Assessment Worksheets** Proj. No. ____	Prepared By ____ Checked By ____ Sheet ___ of ___ Page ___ of ___

WORKSHEET **4** — **PRODUCTS SUMMARY**

Attribute	Description		
	Stream No. ____	Stream No. ____	Stream No. ____
Name/ID			
Component/Attribute of Concern			
Annual Production Rate			
Overall			
Component(s) of Concern			
Annual Revenues, $ ________			
Shipping Mode			
Shipping Container Size & Type			
Onsite Storage Mode			
Containers Returnable (Y/N)			
Shelf Life			
Rework Possible (Y/N)			
Customer Would			
— relax specification (Y/N)			
— accept larger containers (Y/N)			

Appendix VII.1 EPA Pollution Prevention Worksheets (Continued)

Firm ____________ Site ____________ Date ____________	**Pollution Prevention Assessment Worksheets** Proj. No. ____________	Prepared By ____________ Checked By ____________ Sheet ___ of ___ Page ___ of ___

WORKSHEET
5

WASTE STREAM SUMMARY

Attribute	Description		
	Stream No. ____	Stream No. ____	Stream No. ____
Waste ID/Name:			
Source/Origin			
Component or Property of Concern			
Annual Generation Rate (units ________)			
Overall			
Component(s) of Concern			
Cost of Disposal			
Unit Cost ($ per: ________)			
Overall (per year)			
Method of Management[1]			

Priority Rating Criteria[2]	Relative Wt. (W)	Rating (R)	R x W	Rating (R)	R x W	Rating (R)	R x W
Regulatory Compliance							
Treatment/Disposal Cost							
Potential Liability							
Waste Quantity Generated							
Waste Hazard							
Safety Hazard							
Minimization Potential							
Potential to Remove Bottleneck							
Potential By-product Recovery							
Sum of Priority Rating Scores		Σ(RxW)		Σ(RxW)		Σ(RxW)	
Priority Rank							

Notes: 1. For example, sanitary landfill, hazardous waste landfill, on-site recycle, incineration, combustion with heat recovery, distillation, dewatering, etc.

2. Rate each stream in each category on a scale from 0 (none) to 10 (high).

Appendix VII.1 EPA Pollution Prevention Worksheets (Continued)

Firm ______ Site ______ Date ______	**Pollution Prevention Assessment Worksheets** Proj. No. ______	Prepared By ______ Checked By ______ Sheet ___ of ___ Page ___ of ___

WORKSHEET 6

OPTION GENERATION

Meeting format (e.g., brainstorming, nominal group technique) ______

Meeting Coordinator ______

Meeting Participants ______

List Suggestion Options	Rationale/Remarks on Option

Appendix VII.1 EPA Pollution Prevention Worksheets (Continued)

Firm ____________ Site ____________ Date ____________	**Pollution Prevention Assessment Worksheets** Proj. No. ____________	Prepared By ____________ Checked By ____________ Sheet ___ of ___ Page ___ of ___

WORKSHEET
7

OPTION DESCRIPTION

Option Name: ____________

Briefly describe the option: ____________

Waste Stream(s) Affected: ____________

Input Material(s) Affected: ____________

Product(s) Affected: ____________

Indicate Type:

☐ Source Reduction
- _____ Equipment-Related Change
- _____ Personnel/Procedure-Related Change
- _____ Materials-Related Change

☐ Recycling/Reuse
- _____ Onsite
- _____ Offsite
- _____ Material reused for original purpose
- _____ Material used for a lower-quality purpose
- _____ Material sold

Originally proposed by: ____________ Date: ____________

Reviewed by: ____________ Date: ____________

Approved for study? _______ yes _______ no By: ____________

Reason for Acceptance or Rejection ____________

Appendix VII.1 EPA Pollution Prevention Worksheets (Continued)

Firm ____ Site ____ Date ____	**Pollution Prevention Assessment Worksheets** Proj. No. ____	Prepared By ____ Checked By ____ Sheet ___ of ___ Page ___ of ___

WORKSHEET **8**

PROFITABILITY

Capital Costs

Purchased Equipment ____

Materials ____

Installation ____

Utility Connections ____

Engineering ____

Start-up and Training ____

Other Capital Costs ____

Total Capital Costs ____

Incremental Annual Operating Costs

Change in Disposal Costs ____

Change in Raw Material Costs ____

Change in Other Costs ____

Annual Net Operating Cost Savings ____

$$\text{Payback Period (in years)} = \frac{\text{Total Capital Costs}}{\text{Annual Net Operating Cost Savings}} = \_\_\_\_$$

Appendix VII.2 CAP Management Categories

Recovery

Metals Recovery

M011[a]	High temperature metals recovery
M012	Retorting
M013	Secondary smelting
M014	Other metals recovery for reuse: e.g., ion exchange, reverse osmosis, acid leaching
M019	Metals recovery — type unknown

Inorganics Recovery

M031	Acid regeneration
M039	Other recovery — type unknown

Organics Recovery

M021	Fractionation/distillation
M022	Thin film evaporation
M023	Solvent extraction
M024	Other solvent recovery
M029	Solvents recovery — type unknown
M032	Other recovery: e.g., waste oil recovery, nonsolvent organics recovery

Energy Recovery — Liquids

M051	Energy recovery — liquids
M059	Energy recovery — type unknown

Energy Recovery — Sludges/Solids

M052	Energy recovery — sludges
M053	Energy recovery — solids
M059	Energy recovery — type unknown

Treatment

Stabilization/Chemical Fixation

M111	Stabilization/chemical fixation using cementitious and/or pozzolanic materials
M112	Other stabilization
M119	Stabilization — type unknown

[a] System Type codes as defined in U.S. Environmental Protection Agency, *1991 Hazardous Waste Report Instructions and Forms,* EPA Form 8700-13A/B (5-80) (Revised 08-91), OMB #2050-0024, pp. 90–91.

Appendix VII.2 CAP Management Categories (Continued)

Incineration — Liquids/Gases	
M041	Incineration — liquids
M044	Incineration — gases
M049	Incineration — type unknown
Incineration — Sludges/Solids	
M042	Incineration — sludges
M043	Incineration — solids
M049	Incineration — type unknown
Fuel Blending	
M061	Fuel blending
Hazardous Wastewaters and Sludges Treatment	
M071	Chrome reduction followed by chemical precipitation
M072	Cyanide destruction followed by chemical precipitation
M073	Cyanide destruction only
M074	Chemical oxidation followed by chemical precipitation
M075	Chemical oxidation only
M076	Wet air oxidation
M077	Chemical precipitation
M078	Other aqueous inorganic treatment: e.g., ion exchange, reverse osmosis
M079	Aqueous inorganic treatment — type unknown
M081	Biological treatment
M082	Carbon adsorption
M083	Air/steam stripping
M084	Wet air oxidation
M085	Other aqueous organic treatment
M089	Aqueous organic treatment — type unknown
M091	Chemical precipitation in combination with biological treatment
M092	Chemical precipitation in combination with carbon adsorption
M093	Wet air oxidation
M094	Other organic/inorganic treatment
M099	Aqueous organic and inorganic treatment — type unknown
M101	Sludge dewatering
M102	Addition of excess lime
M103	Absorption/adsorption
M104	Solvent extraction
M109	Sludge treatment — type unknown
M121	Neutralization only
M122	Evaporation only
M123	Settling/clarification only

Appendix VII.2 CAP Management Categories (Continued)

M124	Phase separation (e.g., emulsion breaking, filtration) only
M125	Other treatment
M129	Other treatment — type unknown
	Disposal
Landfill	
M132	Landfill
M133	Surface impoundment (to be closed as a landfill)
M137	Other disposal
Deepwell/Underground Injection	
M134	Deepwell/underground injection
M137	Other disposal
Land Treatment/Farming	
M131	Land treatment/application/farming
M137	Other disposal
	Transfer/Storage
Transfer/Storage	
M141	Transfer facility storage, waste was shipped off-site with no on-site treatment, disposal, or recycling (TDR) activity

From U.S. Environmental Protection Agency, *Guidance for Capacity Assurance Planning — Capacity Planning Pursuant to CERCLA §104(c)(9),* U.S. EPA, Washington, D.C., 1993.

8 Site Correction and Remedial Planning

The primary objective of hazardous waste management is the handling of waste in a manner that prevents harm to public health and the environment. Whatever the degree of success in achieving this goal, the fact remains that large numbers of sites have been contaminated with hazardous wastes.

When enacted in 1980, the Comprehensive Environmental Response, Compensation, and Liability Act (CERCLA) was designed to respond to situations involving the past disposal of hazardous substances. Chapter 1 illustrates how costly it is to clean up a CERCLA contaminated site.

CERCLA complements the 1976 Resource Conservation and Recovery Act (RCRA), which regulates ongoing hazardous waste handling and disposal. The RCRA amendment of 1984, called the Hazardous and Solid Waste Amendments (HSWA), provides a framework to address contamination at currently operating treatment, storage, and disposal (TSD) facilities. Under HSWA, all TSD facilities must undertake efforts to investigate the level of contamination and to conduct cleanup of contamination at, or from, the facility, including releases from past disposal.

This framework is known as the RCRA correction action program, and the sites are termed "RCRA corrective action" sites. Under RCRA provisions, corrective action can be initiated either through the permitting process, which requires the facility to obtain a permit for performing corrective action, or through a compliance or consent order compelling the facility to conduct a corrective action.

Statistically, current TSD facilities are composed of approximately 4700 land disposal, incinerator, and treatment and storage facilities. Of this total, about 3700 facilities may need corrective action. The data indicate that the number of TSD facilities that may ultimately need corrective action could be three times the number currently on the Superfund National Priorities List (U.S. EPA, 1990).[175] This massive number of RCRA corrective action sites will obviously elevate the workload of the program to a level that can be critical in terms of the overall management of RCRA permitting.

While RCRA provides a mechanism to deal with hazardous waste contaminated sites at TSD facilities, no specific requirement is currently in place to address RCRA contaminated sites found in a business that generates hazardous waste other than stating that it is illegal to have a disposal/storage activity without RCRA permit. The provision suggests that contaminating a site is seen as an illegal act of storing and/or disposing of a waste. This chapter defines these sites as "RCRA contaminated sites."

The total number of TSD facilities — 4700 — is relatively small compared to the number of hazardous waste generators. In 1989, the number of large-quantity generators (LQGs) was estimated to be 20,233 (U.S. EPA, 1993).[176] The comparison does not suggest that the number of RCRA contaminated sites will surpass the number of RCRA corrective action sites. As no data are currently available, it is sufficient to say that the number of RCRA contaminated sites is large enough for serious policy planning consideration.

As Von Oppenfeld (1993)[177] reveals, there is little doubt, based on the past disposal practices documented in many audits and due diligence reports, that hazardous wastes were intentionally or accidentally placed on the land during the 1980s at various businesses. Accidental spills and leaks may happen during facility operation or during the 90-day storage. In many instances, the leaks were discovered more than 90 days after the accident; consequently, such a spill or leak can be interpreted as illegal storage or disposal of hazardous waste. Von Oppenfeld argues that even if the contention were to be made that the hazardous waste placement on the ground was not intended for disposal, and that there is an intent to eventually place such waste in an appropriately permitted off-site facility, such activity still requires some sort of TSD permit.[178]

Depending on their regulatory status, hazardous waste contaminated sites can be addressed from a number of approaches. Three alternative approaches, or site remediation categories, are briefly introduced in the preceding paragraphs. The categories are (1) CERCLA sites, (2) RCRA corrective action sites, and (3) RCRA contaminated sites. Each of these three sites is discussed in detail in the next sections.

CERCLA SITE REMEDIATION

Statutory provisions in CERCLA are designed to address problems and issues associated with sites contaminated with hazardous substances and pollutants. Hazardous substances are defined in Section 101(14) of CERCLA. They include "hazardous wastes" defined under RCRA, "hazardous substances" and "toxic pollutants" under the Clean Water Act, "hazardous air pollutants" regulated under the Clean Air Act, and "hazardous chemical substances" under the Toxic Substances Control Act.

One of the important features of CERCLA is the creation of the *hazardous substance fund* to be used by the EPA in cleaning up CERCLA sites. This fund, known as "Superfund," can be used to pay the EPA's cleanup and enforcement costs, certain natural resource damages, and certain claims of private parties. An example of the latter is a request for reimbursement when a consultant who has performed a cleanup has been unable to obtain payment from the site owner.

There are four major provisions that characterize CERCLA. First, the statute provides the EPA with the **authority to clean up** CERCLA sites, and it lists the procedures and standards that must be followed in remediating the sites. The National Contingency Plan (NCP) represents the primary document for CERCLA response action. It sets forth the procedures that must be followed by the EPA and by private parties in performing CERCLA response actions.

The National Priorities List (NPL), which is part of the NCP, prioritizes CERCLA cleanup sites based on the levels of risks to public health, welfare, or the environment. The list must be updated annually. The priority listing of CERCLA sites takes into consideration a variety of factors, including the extent of population at risk, the hazard potential, the potential for drinking-water supplies contamination, and the threat to ambient air because of the existence of a site.

Second, the **enforcement provisions** of CERCLA identify the classes of parties liable under the statute, detail the provisions of cost recovery, and provide guidance on cleanup settlements with the EPA. CERCLA defines the parties responsible for the contamination of a site to include past and present owners or operators of the site, generators of hazardous waste found at the site, and transporters of hazardous waste to the site. Once the potentially responsible parties (PRPs) are identified, the EPA can negotiate a legal agreement or unilaterally order them to take remedial action responsibility of the site. Parallel to the CERCLA goal of acquiring the PRPs to take appropriate response action to clean up a contaminated site, we learned in Chapter 3 that CERCLA places "strict, joint, and several" liabilities on responsible parties associated with the site.

Liability provisions under CERCLA include:

1. allowing the EPA and private parties to recover their cleanup costs;
2. permitting the EPA to seek a judicial order to require a liable party to abate an endangerment to human health, welfare, or the environment;
3. permitting the EPA to take administrative actions to compel private parties to undertake actions to protect public health, welfare, or the environment;
4. allowing private parties to bring citizen suits to enforce CERCLA's provisions; and
5. providing authority for the natural resource trustee to bring actions for damages of natural resources.

In order to recover cost, it is necessary that the claimant justifies that the cost is associated with *response* activities, which consist of *removal* and *remedial* actions. Examples of recoverable costs include those associated with sampling, services of environmental consultants and attorneys, and the identification and disposal of hazardous substances. Examples of unrecoverable costs are medical monitoring costs, lost profits, and general damages.

Although some CERCLA cases have proceeded through trial, normally the dispute is addressed through settlement. The EPA prefers settlement because it conserves Superfund monies and the EPA's resources. PRPs prefer this alternative because it allows them to exercise greater control over the planning and implementation of remedial actions.

In the third provision, CERCLA specifies when releases of hazardous substances must be reported. The statute does not specify the amount of a hazardous substance that must be present before a response action can be taken or before a party can be found liable for the release. However, CERCLA's **reporting requirements** mandate reporting a release of hazardous substances only when a specified reportable quantity is released. This reporting requirement is discussed in Chapter 9.

CERCLA provides authority for the EPA to respond to a release or a substantial threat of a release of any contaminant that may present an imminent and substantial danger to human health or welfare. The term contaminant or pollutant encompasses just about anything. While the EPA can respond to and clean up a site polluted by a contaminant, CERCLA does not authorize the EPA to recover its cleanup costs from private parties or to issue an order compelling the parties to conduct a cleanup when the contaminant is not a hazardous substance. While the definition of a contaminant is broad, this breadth of coverage has no practical impact on private parties (Lee, 1993).[179]

In the fourth provision, CERCLA contains procedures to be followed for the cleanup of **federal facilities**. The greatest environmental concerns in this case are for those associated with the military facilities (Department of Defense and Department of Energy) that were constructed in response to World War II and the Cold War. Past disposal practices that contribute to contamination at these facilities include the use of unlined pits, discharge to the ground, and on-site burning of wastes.

The 1986 Superfund Amendments and Reauthorization Act (SARA) requires federal facilities to comply to the same extent as private entities. These include, for example, requirements related to site assessments, hazardous ranking, and evaluation procedures. SARA also establishes requirements that are unique to federal facilities, such as the creation of a Federal Agency Hazardous Waste Compliance Docket, which lists facilities that manage hazardous waste or have potential hazardous waste problems. The list is used to determine whether a facility should be placed on the NPL.

The following two principal operations represent CERCLA's remedial action framework:

- The EPA may conduct the cleanup itself and seek to recover the cleanup costs from PRPs in a subsequent cost-recovery effort.
- The EPA may compel the PRPs to perform the cleanup through administrative and judicial proceedings.

In terms of implementation, CERCLA sets forth two categories of response actions. They are removal and remedial actions. **Removal actions** are conducted promptly in response to environmental emergencies to eliminate or minimize the threat of a hazardous waste site. Examples of removal actions include the immediate cleanup of hazardous waste spilled from a container, the installation of a fence around a hazardous waste site, and the provision of alternate water supplies for a local community.

Removal actions can occur as part of the initial action to an extensive remedial action at an NPL site. Removals can also occur at a site not listed on the NPL. Any removal action must be completed within one year and at a cost of less than $2 million, unless there are reasons for extending the time and the cost — such as, the action is required as part of a larger approved remedial action, or that there is an immediate risk to public health or the environment.

Remedial actions are long-term, permanent cleanups to eliminate any threat that a site may pose. Only sites listed on the NPL qualify for remedial actions. Examples of remedial actions are excavations, installation of clay covers, and neutralization or destruction of hazardous substances. Because remedial action is a complex, lengthy, and costly process, detailed and comprehensive requirements are established by CERCLA. Major components of this process are discussed in the following section.

Selecting Candidate Sites

This process starts with listing the sites with releases or threatened releases in the Comprehensive Environmental Response, Compensation, and Liability Information System (CERCLIS). This data base represents the official national inventory of CERCLA sites. The list is compiled by the EPA from various information sources, such as CERCLA's reporting, submissions from states, and citizen complaints. Once it is compiled, the EPA performs a **preliminary assessment** (PA) by reviewing existing site data and, if necessary, conducting a site inspection. The goal is to screen the sites by determining whether each site presents a risk of release of hazardous substances.

If the PA indicates that a site presents a threat and will probably score high enough to be listed on the NPL, the EPA will undertake additional **site investigations** to obtain more information about the hazardous substances at the site, the movement (or *migration)* of the contaminants, and the potential exposure to humans and the environment. All of the information collected is used to score the site in accordance with the NPL's **hazard ranking system** (HRS).

The HRS utilizes a variety of scoring factors, such as the toxicity of the substances, the location of the population at risk, exposure pathways, and the threats to the human food chain, ambient air, and groundwaters. The scoring system runs from 1 to 100 with a 100 score representing the worst site. A site will be listed on the NPL if it scores above 28.5. Once it is listed on the NPL, CERCLA's remedial action is committed to the listed sites.

Scoping the Remedial Action

The scoping of remedial activities is determined from the result of the Remedial Investigation/Feasibility Study (RI/FS). The RI/FS assesses the conditions of a site and examines alternatives to remedy the site. The purpose of the **remedial investigation** (RI) is to evaluate the nature and extent of releases of hazardous substances and to determine those areas of a site where releases have created damage or the threat of damage to public health or the environment. The

RI requires extensive soil and groundwater sampling and voluminous reports describing the detailed remedial investigations. These investigations examine the nature of the site's geology and hydrogeology, identify the sources, types, and mobility of contamination, and determine the nature of the threat to public health or the environment.

The conclusions of the RI provide valuable documentation and information on the sources of contamination, the nature and extent of contamination, and the actual and potential exposure pathways. The RI finding provides the EPA with the basis for designing a **feasibility study** (FS) of the site. The purpose of the FS is to develop a range of remedial action alternatives to address the contaminated site. Of the options evaluated, the EPA will select the best alternative as the remedy of the site.

The entire RI/FS process can take years to complete. It is time-consuming and very costly. For efficiency consideration, the RI/FS recommendations may include phasing the cleanup effort by dividing the site into some **operable units**. A site may consist of an operable unit that will address the remediation of groundwater at the site, and another operable unit that is designed to clean up and isolate the sources of contamination. Using this approach, it is possible to remediate certain portions of the site, while other portions are undergoing further study and evaluation.

In remedying the site, one of the most controversial issues at any CERCLA site is the level of cleanup that must be achieved before the site can be considered "clean." The underlying issue is whether the site must be cleaned up to the predisposal conditions, or, if not, what levels of cleanup are considered adequate.

Determining the Cleanup Level

The significant issues regarding the cleanup standard in the CERCLA program are the required level of groundwater remediation and the level of residual soil contamination that is allowed to remain at the site. The decision on these two issues determines the type of remedial technology that can be used, and the extent of required excavation at the contaminated site. Consequently, this decision will affect the remedial cost; therefore, all parties with a stake in the CERCLA site have serious concerns about how cleanup standards are determined.

CERCLA provides a solid preference for remedies that are permanent and that involve the treatment of hazardous substances to reduce their volume, toxicity, and mobility. Remedies that are designed to contain contamination within a site are not favored as well as the off-site transport and disposal of untreated hazardous substances.

The statute requires that a selected remedy achieve all *"applicable or relevant and appropriate requirements"* (ARARs). ARARs for hazardous substances and contaminants are standards, criteria, requirements, or limitations imposed under any federal environmental law or under a state environmental or facility siting law if it is more stringent than the federal standard. The application of ARARs at a CERCLA site implies that the selected remedies must achieve the highest cleanup level, which in practice is extremely conservative and costly. CERCLA does not specify which ARARs apply to a specific site. The selection of ARARs represents the most important part of the RI/FS process.

The NCP provides more detailed criteria for selecting remedies and applying ARARs (40 CFR § 300.430[f]). The first set of criteria is called the *threshold criteria,* which require the remedy (1) to provide overall protection of human health and the environment, and (2) to comply with ARARs. If a proposed remedy fails to meet both criteria, it will not be selected.

If the proposed remedy meets the threshold criteria, it will be further screened by the *primary balancing criteria,* which consist of the following measures:

1. long-term effectiveness and permanence;
2. reduction (through treatment) of volume, toxicity, and mobility;
3. short-term effectiveness;
4. implementability; and
5. cost.

Balancing the first three measures with the fifth criterion provides the planner with a mechanism to check if a proposed remedy is, in fact, cost-effective.

The last screening tool is the *modifying criteria,* which refer to the state acceptance and community acceptance. Normally, if a proposed remedy passes the threshold and primary balancing criteria, it also meets the modifying criteria.

The NCP expands the discussion of ARARs in 40 CFR § 300.400(g) by providing the definitions of "applicable" and "relevant and appropriate" requirements. An **applicable** requirement is defined as one that specifically addresses a hazardous substance, contaminant, or pollutant; remedial action; location; or other circumstance found at a CERCLA site. If a requirement is not applicable to a specific release or remedial action, the requirement may still be **relevant and appropriate**.

In evaluating relevance and appropriateness, the NCP lists eight comparisons that must be examined to determine whether a requirement addresses problems or situations sufficiently similar to the circumstances of the release or proposed remedial action, and whether the requirement is well-suited to the site and, therefore, is both relevant and appropriate. The comparison factors are:

1. the purpose of the requirement and the purpose of the CERCLA action;
2. the medium affected by the requirement and the medium contaminated at the CERCLA site;
3. the substances regulated by the standard and the substances found at the site;
4. the activities regulated by the requirement and the activities contemplated at the CERCLA site;
5. any variances, waivers, or exemptions of the requirement, and their applicability to the site;

6. the type of place regulated and the type of place affected by the release or CERCLA action;
7. the type and size of the facility regulated and the type and size of the facility affected by the release or contemplated by the CERCLA action; and
8. any consideration of the use of the affected resources in the requirement and the potential use of the affected resources at the CERCLA site.

Once the selection of ARARs is completed and the RI/FS is finalized, the EPA issues a Record of Decision (ROD). The ROD contains all of the facts, analyses, and policy determinations. It explains how the selected remedy is protective of human health and the environment, how the ARARs will be achieved, why the remedy is cost-effective, and why it represents permanent solutions. The ROD must respond to public comments regarding the selection of the remedy. The ROD must be placed in the site's administrative record. This record is critical because it will be used in any judicial proceedings.

The issuance of the ROD is followed by the completion of a **remedial design** (RD). Through an RD, the selected remedy — which is recorded in the ROD — is translated into a detailed design for construction and operation. The construction and operation phase is called the **remedial action** (RA) of a CERCLA site.

The mission and procedure of remedial planning for RCRA sites are somewhat different from those for CERCLA sites. The following section discusses planning issues related to sites contaminated with RCRA hazardous wastes. The discussion focuses on contamination occurring at hazardous waste TSD facilities that are still in operation.

RCRA CORRECTIVE ACTION

Prior to the 1984 HSWA, the EPA had to rely on RCRA's Section 7003 enforcement authority to require TSD facilities to clean up releases of hazardous waste or constituents that resulted from facility operations. **The 7003 order** is used to order cleanup of an "imminent and substantial endangerment to health or the environment" caused by the handling of solid waste and hazardous waste. Similarly, under Section 3013 of RCRA, the EPA could issue **the 3013 order** to require facilities to conduct investigations when the presence of hazardous waste or releases of hazardous waste might present a substantial hazard to human health or the environment.

The 1984 HSWA authorizes the EPA to force TSD facilities to perform corrective action for releases from a facility. Under **Section 3004(u)**, the EPA must require corrective action for all releases of hazardous waste or constituents from any *solid waste management unit* (SWMU) at a TSD facility seeking an RCRA permit, *regardless of the time* at which the waste was placed in such unit. The term SWMU includes, for example, containers, tanks, drum storage areas, accumulation areas, surface impoundments, waste piles, land treatment units, landfills, sumps, incinerators, and underground injection wells. The 3004(u) requirement applies to facilities seeking ordinary operating permits as well as those seeking closure or post-closure permits. This provision also requires that the TSD facility demonstrate *financial assurance* for any corrective action taken.

Section 3008(u) does not require proof of some threat of endangerment before the EPA can require corrective action for a hazardous waste release from an SWMU. In practical terms, it would be arbitrary and capricious for the EPA to require extensive corrective action where no threat to human health or the environment existed (Davis, 1993).[180] Davis argues that, in fact, the regulation implementing this statutory provision (i.e., 40 CFR 264.101) states that corrective action will be required as necessary to protect human health and the environment.

Section 3008(v) directs the EPA to issue regulations requiring TSD facilities to undertake corrective action *beyond the facility boundary* where necessary to protect human health or the environment. This requirement will not apply if the facility can demonstrate that, despite best efforts, it was unable to obtain permission from other property owners to conduct such action. The facility is still required to take on-site measures to prevent off-site releases. Until such regulations are promulgated, corrective action orders can be issued to require necessary corrective action.

Under **Section 3008(h)**, the EPA is authorized to compel owners and operators of *interim status* TSD facilities to conduct corrective action when there is or has been a release of hazardous waste from such facility. The 3008(h) order is issued when it is necessary to protect human health or the environment.

The authority granted by HSWA clearly indicates that both the permitted TSDs and the interim status facilities are subject to corrective action requirements. Corrective action can be initiated within the RCRA program, either through the permitting process — which requires the facility to obtain a permit for performing corrective action — or through a compliance or consent order compelling the facility to conduct a corrective action. Both processes are essentially alike, except that orders do not require a public comment procedure.

RCRA corrective action is a three-step program consisting of the following (McCarroll, 1993):[181]

1. Site characterization identifies the contamination level of the site. This step consists of an **RCRA Facility Assessment** (RFA) and an **RCRA Facility Investigation** (RFI). Conducting an RFA is the responsibility of the regulatory agency, while completing an RFI is the responsibility of the TSD facility.
2. Remedy selection describes how the facility should clean up the site. This step, which is known as **Corrective Measures Studies** (CMS), starts with determining the need for remedial measures and continues with selecting appropriate remedies, including the "no action" alternative. The CMS is the responsibility of the facility.
3. Site remediation is the responsibility of the facility and is defined as the **Corrective Measures Implementation** (CMI). CMI includes the design, construction, operation, and maintenance of the selected remedies. This step requires active regulatory agency oversight.

In addition to those three steps, **interim measures** can be implemented at any time to respond to immediate threats. Interim measures are the individual actions to stabilize site conditions. The goals are to control or abate immediate threats to human health or the environment, to prevent or minimize the spread of contaminants, and to immediately initiate cleanup actions. Examples of interim measures are installing a landfill cap to reduce infiltration of surface water into a landfill, installing a gas collection system to intercept migrating landfill gas, removing contaminated surface soils, and shutting down a contaminated water supply well.

The implementation scheme of the corrective action provisions described in the preceding paragraphs is governed by the EPA's National RCRA Corrective Action Strategy, which was announced by the agency in 1985. On July 27, 1990, the EPA proposed a new comprehensive regulatory framework for implementing the corrective action program. The new framework, which will be contained in Subpart S of 40 CFR Part 264, essentially codifies many of the provisions currently found in the strategy materials.

Because of the importance of site characterization in the context of hazardous waste planning, the next sections discuss this activity in detail. As components of site characterization, both the RFA and the RFI are elaborated in terms of goals, expected outputs, and processes.

RCRA Facility Assessment

The purpose of the RFA is to identify and gather information on releases of pollutants and make preliminary determinations about releases, the need for corrective measures, further investigations, and interim measures. It also screens those units in a facility that do not pose a threat to human health and the environment. The overall process will furnish a description of the facility in terms of its:

- operation, permitting history, and the characteristics of the **units**;
- **waste** generation and management;
- past and present actual and potential **releases** of hazardous waste or constituents to all environmental media;
- probable contaminant movement, or **migration**, through five environmental pathways: groundwater, surface water, air, soil, and subsurface gas (e.g., methane gas at landfill);
- likely human receptors that may be or have been **exposed** or affected by those actual or potential releases; and
- **need** for corrective measures, interim measures, and further investigations.

Although the RFA addresses all releases of pollutants, some releases are lawful if the concentration level of the pollutants is below acceptable limits. Releases that cause concern are those that have a pollution level above standards.

In order to assess releases of concern, the RFA provides a pre-structured evaluation of release potential through a four-step process:

1. preliminary review (PR),
2. visual site investigation (VSI),
3. sample collection (if necessary), and
4. corrective action need assessment as specified in the RFA report.

The **preliminary review** (PR) may be the most time-consuming aspect of the RFA process; however, a thorough review will pay major dividends in reduced time and effort required to complete the RFA (Corbaley, 1993).[182] In terms of time frame, the typical RFA can be conducted within 16 to 21 weeks (Table 8.1).

The PR provides a clear understanding of the facility, focuses the subsequent visual site investigation, and identifies information gaps early in the RFA. Information needed during the PR includes a facility map, a topographic map, a copy of the RCRA permit application, the facility hydrogeological study, closure plans, spill reports, biennial reports, inspection reports, enforcement orders, county soil survey, and aerial photos.

The PR report describes the facility, permit history, waste management practice, environmental setting, facility units, and data gaps in terms of the need of additional information specific to each unit. These data gaps will be filled during the **visual site investigation** (VSI). The data needed are not necessarily about the releases.

In addition to filling in data gaps identified during the PR process, the VSI entails the on-site collection of visual evidence of the facility units, current hazardous waste management practices, pollution controls, releases, and routes of exposure. The VSI includes interviews with facility personnel. During an interview, Wesling (1993)[183] suggests asking all questions necessary to fill data gaps and to talk with not only the environmental manager, but also the employees involved in the day-to-day operations. If necessary, request that facility personnel repeat information or slow the pace of the VSI in order to take complete notes.

Although the VSI is intended to fill in data gaps, the gaps may still exist even after the completion of the VSI. In this instance, field data and environmental **sampling** may be required. As previously noted, the regulatory agency is the responsible party in conducting the RFA; however, environmental sampling as part of the RFA can be implemented by requiring the facility to do it pursuant to a permit condition.

Once environmental sampling is completed, all information obtained from the PR, VSI, and sampling are analyzed and described in the **RFA report**. The report consists of two major parts: the non-confidential information and the confidential information. The first part describes the purpose of the RFA, facility description, permit history, waste management practice, environmental setting, facility units, and exposure pathways. The confidential part of the report is the summary and recommendations for corrective action (Corbaley, 1993).[184] Based on the characterization of releases and the potential for exposure to humans and

Table 8.1 Typical RFA Schedule

Activity	Number of days
File review	5
Draft PR report	30–45
Draft review	7
Final PR report	7
Notify TSD facility, allow the facility to gather requested information	21–28
Visual site investigation	1–7
Draft RFA report	30–40
Draft review	7
Final RFA report	7
Overall RFA process	115–153

the environment, the possible next steps in the corrective action process can be (1) interim measures, (2) a referral to the appropriate program, such as Superfund, air program, or NPDES, or (3) the initiation of RCRA Facility Investigation.

RCRA Facility Investigation

The goal of an RCRA Facility Investigation (RFI) is to identify the contamination problem in a TSD facility and examine how bad the problem is. Specific information pertinent for the RFI includes the description of the facility problems, the types of contaminants released, the sources of contaminants, the pathways, the extent of contaminant migration, and the threats to human health or the environment.

The RFI is the responsibility of the respective TSD facility. There are five elements that comprise the RFI. They are:

- examination of existing condition,
- development of the RFI workplan,
- evaluation of the RFI workplan,
- health and safety plan, and
- public involvement plan.

When conducting an *assessment of current conditions,* the investigation must focus on visual observation of the facility, the facility's compliance history, the environmental setting, and the existing level of contamination. Once a conceptual model of contaminant migration is developed, the RFI examines the impacts of contamination based on the model, evaluates possible interim measures, and identifies the need for additional data.

The **RFI workplan** elaborates the overall management of the process and details the field investigation plan, sampling plan, quality assurance and quality control (QA/QC), and data management. The procedures for sample collection include a map showing sample locations, personnel and equipment requirements, equipment calibration, equipment decontamination procedures, disposal of contaminated materials, sample packaging and shipment, and sampling documentation. For a large or complex site, phasing the RFI into more than one workplan is preferred, prioritizing the worst areas first (Saracino, 1993).[185]

Once the RFI workplan is completed, the TSD facility must obtain approval from the environmental agency. The agency will *review the workplan* with the assumption that what is written on the workplan may not necessarily happen in the field. When reviewing the plan, the agency will ensure that all areas are adequately covered and that the sampling plan is designed to look for the right contaminants in the right places. Site visits and audit sampling procedures are necessary; the agency will take additional samples during site visits.

In addition to the workplan, the RFI report also contains a health and safety plan and a public involvement plan. The purpose of the **health and safety plan** is to ensure safety in all aspects of the project. The plan should include an assessment of hazards and preventive measures, a list of and how to access the personal protection equipment necessary for the site, site organization, and a list of names and phone numbers of emergency contacts.

A **public involvement plan** is not required, but is recommended. The goal of the plan is to communicate to the public the objectives of the project, methods to achieve those objectives, and the status of the project. When developing the plan, ensure that the public's concerns are identified and addressed.

The completion of the RFI report indicates the end of the site characterization phase of the RCRA corrective action; remedy selection and site remediation phases can subsequently be implemented. These three phases of RCRA corrective action are designed for remediating a site in which a TSD facility is currently in operation. The next section addresses the remediation of an RCRA contaminated site, i.e., a contaminated site in which a hazardous waste generator is conducting its regular business.

RCRA CONTAMINATED SITE REMEDIATION

Remediating an RCRA contaminated site can be problematic in a number of ways. First, before any decision is made on remediating an RCRA site, it is imperative that the regulatory status of contaminants is determined. For example, when a hazardous waste is found as a contaminant, the site may not necessarily be an RCRA contaminated site if the contamination occurred prior to 1980. Additionally, when a hazardous waste commingles with another hazardous substance regulated under CERCLA, the mix of contaminants under more than one statute will certainly create a policy decision issue.

There are a number of statutes addressing contaminated sites, such as CERCLA and RCRA Subtitles C, D, and I. Although each statute refers to specific regulated contaminants, cleanup requirements for each medium (soils, water, and air) are not consistently provided. Basically, there is no single statute that comprehensively addresses multimedia contamination. In reality, multimedia contamination is normal and is usually found at most RCRA and CERCLA contaminated sites.

As a consequence of differing statutory requirements, the level of cleanup is also different among statutes. There are three options regarding the cleanup level required by current laws. They are (1) no-detect level, (2) background level, and (3) risk-based or health-based level.

The no-detect option is the most stringent requirement, because it does not allow the existence of any suspected contaminant at a remediated site. The background option requires the cleanup efforts to be conducted with the goal of reducing the contamination of each suspected contaminant to a level similar to the area surrounding the site. The risk-based level requires the cleanup efforts to reach a level that is acceptable based on risk assessment criteria. The criteria are contaminant-specific and site-specific. Depending on the nature of contamination, the risk-based level can be higher or lower than the background level.

In reality, all of these requirements may not be financially feasible. Remediating a site even to the highest of those three levels can be very costly. This may result in the containment of the site by capping it with a notice to the deed to inform the public about the contamination status of the site. Even worse than containment of the site is the possibility of closing the business operating on the property and leaving the site as-is, until the site is included in the federal or state Superfund program.

Clearly, the combination of the above issues results in the need to address RCRA contaminated sites as comprehensively as possible by integrating the regulatory, technical, and managerial aspects into the decision-making process. The following paragraphs examine an integrated model that can be used as an example to address RCRA contaminated sites.

A Comprehensive Approach to RCRA Contaminated Sites

An active RCRA contaminated site means that while there is evidence of contamination in the business property, the business is still in operation to produce products and services to customers. In this regard, the active RCRA contaminated site differs from a Superfund site in that it represents a controlled use situation. This unique characteristic provides an opportunity for considering a variety of cleanup mechanisms and cleanup levels appropriate for certain RCRA contaminated sites. Soesilo and Curry (1994)[186] suggest that this opportunity be evaluated in a comprehensive decision-making process (Figure 8.1).

Figure 8.1 shows that the decision-making process starts with determining the regulatory status of the contaminants. A number of possibilities may be discovered during the waste determination process; some wastes may be RCRA wastes, CERCLA wastes, underground storage tank (UST) wastes, or RCRA Subtitle D solid wastes. A site contaminated with RCRA hazardous waste prior to 1980 is considered a CERCLA site. A site contaminated with RCRA hazardous waste after 1980 is considered an RCRA contaminated site. The dividing line between CERCLA and RCRA contaminants is 1980, because the respective contaminants were regulated as RCRA hazardous wastes beginning in 1980.

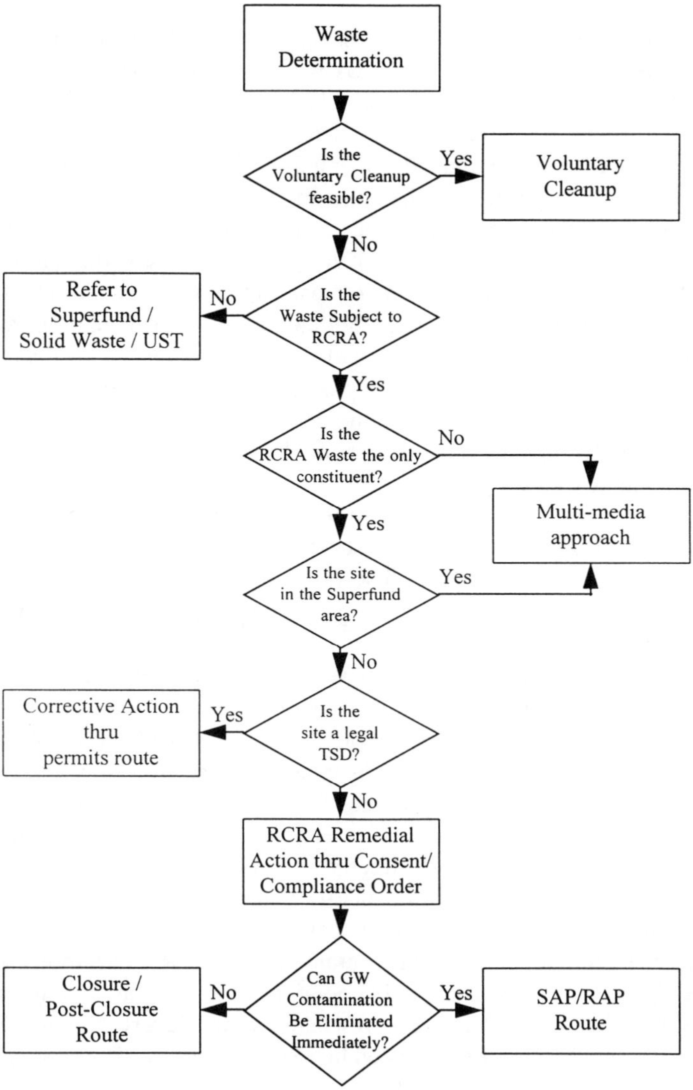

Figure 8.1 RCRA Corrective/Remedial Actions.

A waste that is not listed as RCRA hazardous waste but shows a characteristic of hazardous waste is considered RCRA waste. A site contaminated with petroleum products because of a release from a leaking UST is regulated under the UST program. However, if the release comes from an above-ground tank, the site may be defined as RCRA contaminated site or a Subtitle D solid waste site, depending on the hazardous constituent of the contaminants.

The above discussion shows that determining the regulatory status of the contaminants is significant because it clearly affects the remediation planning of a site. After the properties of the contaminants are known — and if the magnitude of the contamination is found to be "not extensive" — it is appropriate for the owner of the site to perform voluntary cleanup upon approval of the state regulatory agency, if a question exists in the mind of the owner regarding the regulatory requirements of the site.[187] In this case, the owner (or the representative consultant) and the agency must determine that the degree of the contamination warrants voluntary cleanup.

The agency must have proof that the contamination occurred as a result of a discharge of RCRA hazardous waste, CERCLA hazardous substance, UST waste, or Subtitle D solid waste. The remediation planning is eventually performed according to the requirements specified by the respective statute or regulation.

In some states, the UST program does not "approve" voluntary cleanups. Owners or operators are expected to remediate releases expeditiously without prior approval. Power and Leifer (1993)[188] assert that the benefits of facilitating voluntary cleanups, both from an environmental protection standpoint and from the perspective of private parties, are numerous. They argue that voluntary cleanup will result in a far greater number of sites being remediated, many sites will be cleaned up faster and more cost-effectively, the resulting expedited action will decrease the potential for contaminants to migrate, and the cleanup will render property usable and valuable to companies and local communities.

Once it is determined that the contaminant is RCRA hazardous waste, the next question is whether it is the only contaminant or whether it commingles, for example, with hazardous substances regulated under Superfund laws. Similarly, it is also possible that the site is located in a Superfund area. This assessment may result in a situation where the regulatory agency can address the site using RCRA regulations or Superfund provisions.

If the only contaminant is RCRA hazardous waste, the business that is responsible for the contamination can be classified in two ways:

1. The business is a **generator** and is responsible for cleaning up the contamination. The generator is compelled to conduct the cleanup through the signing of an enforcement order issued by the regulatory agency.
2. The business is an **illegal disposal facility** because the contamination is viewed as the result of an act of disposing hazardous waste on-site without a permit.

Depending on how the business owner and the regulatory agency reach a consensus, these two classifications will definitely affect the scope of the cleanup activity. If the site is located on the premises of a TSD facility, the facility is required to obtain an RCRA corrective action permit for the site.[189] If the business is regarded as a generator, the facility may end up conducting the cleanup activities either through the Site Assessment Plan (SAP) and Remedial Action Plan (RAP) route, or through the closure and post-closure route under a formal compliance agreement.

The model illustrates that the flexibility of this approach lies in the fact that there are alternatives for the cleanup of RCRA contaminated sites. The first and the most favorable alternative is the voluntary cleanup option. This option can be used if the remediation can be accomplished "immediately." However, if this is not possible, there are three other options that can also be considered. They are (1) the RCRA corrective action permitting option, (2) the compliance agreement with the SAP/RAP requirement, and (3) the closure and post-closure alternative.

Assessment of the Alternative Cleanup Routes

Among the three options, the **RCRA corrective action permitting** represents the alternative with the most comprehensive requirements. This option is preferred if both soil and groundwater contamination exist. The RCRA corrective action permitting process is specified in the RCRA regulations. Before conducting the cleanup, it is required that the facility develops an approvable closure plan that describes the method and standard used to clean up the contaminated site.

If clean closure is not achievable, i.e., a certain level of contamination is still left in the soil or groundwater, post-closure care must be conducted by the facility. This includes the installing, monitoring, and maintaining of groundwater wells based on an approved post-closure plan. The permitting option requires the facility to provide valid financial assurance to guarantee that a substantial amount of money is available to finance the closure and post-closure activities.

The **SAP/RAP alternative** is a cleanup process which uses the "comparable authority" process (U.S. EPA, 1990)[190] to facilitate cleanup at an RCRA contaminated site. The SAP/RAP does not require financial assurance. The procedure is less comprehensive than the first alternative so that the cleanup activity can be conducted much faster. This alternative is appropriate if no groundwater contamination is detected or if the groundwater contamination can be expeditiously eliminated upon the completion of the cleanup.

The SAP/RAP method and cleanup standard must be approved by the regulatory agency. In essence, the scope and content of a Site Assessment Plan (SAP) is similar to an RCRA Facility Assessment (RFA) and an RCRA Facility Investigation (RFI), while the Remedial Action Plan (RAP) is comparable to the Corrective Measures Studies (CMS) and the Corrective Measures Implementation (CMI) of the RCRA corrective action planning.

The **closure and post-closure** alternative is similar to the RCRA corrective action permitting option in that it requires the facility to perform closure and post-closure based on approved plans. As in RCRA corrective action permitting, the term *closure* means cleaning up the site.

There are two major differences between these options. First, the closure and post-closure alternative is executed through an enforcement action consisting of either a consent or compliance order, not the permitting process. Second, instead of requiring a cleanup procedure as specified in Subpart H of 40 CFR Parts 264 and 265, the order specifies that the facility must provide a cleanup procedure approved by the regulatory agency. This provision is incorporated in the language of the order.

The above discussion reveals that if immediate voluntary cleanup is not possible, the next best option is the SAP/RAP, followed by the closure and post-closure alternative. The RCRA corrective action permitting process represents the last option. The voluntary cleanup and SAP/RAP deal with contamination where clean closure is possible; the closure/post-closure and most RCRA corrective action permitting alternatives are used where clean closure is not achievable, and therefore, post-closure monitoring of the site is necessary. The applicability of this alternative scheme, however, is dependent on the provisions in state laws and regulations.

Regardless of the option used, the issue of "how clean is clean" is still prevalent. In other words, the choice of the cleanup level, whether it is a no-detect, background, or risk-based cleanup standard, is still subject to debate and is normally decided on a case-by-case basis. In addition to the choice of the cleanup level, some other planning issues in remediating a CERCLA site, RCRA corrective action site, or RCRA contaminated site include the streamlining of the process, cost-effectiveness, and the risk to human health or the environment.

ISSUES IN SITE REMEDIATION PLANNING

Site remediation planning is currently being confronted with the intricate relationship between cost, timing, and risk. While the goal of remedial planning is to expedite remediation without ignoring the risk to public health and the environment, the elements of cost and time to implement remediation represent two major barriers to site remediation planning. Stated differently, Travis and Doty (1992)[191] assert that selecting cost-effective remedial alternatives that provide for effective and permanent cleanup, and deciding which sites to remediate, are the two fundamental elements of the decision-making process.

In the early 1990s, the average cost of remediation for a CERCLA site was estimated to be $25 million per site, which was up from $8.5 million two years before. The time required to clean up a CERCLA site can be as long as 10 to 15 years. Costs are rising rapidly, and sites with expenditures of $50 million to $70 million are increasingly common (Ronan III, 1990).[192]

Statistically, more than 33,000 CERCLA sites have been identified as possible *candidates* for listing on the NPL; the candidacy of approximately 19,000 of these sites has been dropped after preliminary assessment (Carr, 1991).[193] Since

1980, more than 1200 sites have been listed in the NPL, and remedial action has been started at over 600 of these sites. Additionally, more than 2600 remedial actions have been performed (Environment Reporter, 1991).[194]

The timing and cost problems are not necessarily particular to CERCLA sites. Similar issues are also applied to RCRA corrective action sites and RCRA contaminated sites. The reason for this type of problem is that each contaminated site is unique, and many have an unusual combination of the technical, regulatory, and political sensitivity problems.

Burke (1992)[195] argues that much of the delay in the remediation process can be attributed to the difficulties of decision-making. The decision-making represents an interface of science and values of those affected by remediation decisions. According to Burke, the interface is most evident in the areas of priority setting, establishing cleanup standards, assessing health risks, and communicating risks to policymakers and the public.

Many analysts believe that more attention should be paid to speeding basic cleanups, and that relatively less attention should be paid to fine-tuning the remedial alternatives or to achieving perfection in site restoration (Russell et al., 1991).[196] In this regard, efforts to streamline the remedial planning and implementation are numerous.

Examples of such effort include the establishment of state Superfund programs, which address CERCLA contaminated sites that fail to be included in the federal Superfund program, and the introduction of the Superfund Accelerated Cleanup Model (SACM). SACM is an EPA pilot project developed in early 1990 to make the federal Superfund process simpler, more flexible, faster, cheaper, and quicker in reducing acute risks. The main SACM features are streamlining the site assessment process, blending the removal and remedial processes, one-stop sampling, using regional decision-making teams, and public involvement throughout the process.

The federal RCRA Implementation Plan (RIP) also discusses the need to prioritize TSD facilities for permits and RCRA corrective action. The RIP recommends that an Environmental Priority Initiative (EPI) ranking system be used. Under this strategy, facilities are scored based on three criteria. The criteria are:

1. environmental significance, such as the type and amount of waste handled and the exposure potential to residences, groundwater, and surface water;
2. environmental benefits due to corrective action, such as creating pollution avoidance and providing incentive for compliance; and
3. other considerations, including compliance history, financial situation, and political sensitivity.

The EPI ranking utilizes a scoring system of 1 to 100, with a 100 score representing the worst RCRA corrective action site. This ranking scheme enables the EPA and the state to prioritize the remediation of RCRA corrective action sites appropriately.

Additionally, the EPA is also initiating the designation of Corrective Action Management Units (CAMUs) within the RCRA corrective action site. CAMU designation is intended to facilitate cleanups by waiving certain requirements for specific RCRA corrective action remedies, based on the assumption that management of remediation wastes is different from management of generated wastes. Because of this assumption, CAMU is only applied for on-site remedial wastes. Moreover, the provision is limited to corrective action under Sections 3004(u), 3008(h), and 7003 of RCRA and is not intended, at least initially, for RCRA closure (Gardner, 1993).[197]

CAMU rules recommend liners when appropriate, and they encourage treatment before placement of the wastes into the unit. Some of the considerations in designating an area into CAMU are:

- The designation will not create unacceptable risks to human health or the environment during waste management.
- The establishment of CAMU will minimize future releases.
- The designation will facilitate the implementation of reliable, effective, protective, and cost-effective remedies.
- CAMU will expedite the timing of cleanup.
- It will enhance long-term effectiveness through the reduction of toxicity, mobility, or volume of wastes.

CAMU rules and the diversity of efforts to expedite the remediation of contaminated sites, combined with the complexity of issues in the remedial decision-making process, certainly support the position that site remediation currently represents a growing and challenging field. This challenge will continue to accelerate as the program matures and many more issues surface for decision-making.

9 Emergency Response and Hazard Management Planning

Hazard management planning is concerned with the safety of people and the environment from potential hazards resulting from hazardous materials accidents, incidents, or improper management of hazardous material. The term *incident* is widely used to refer to a release or spill of material that can be directly or indirectly harmful to human life or damaging to its supporting environment. An incident usually happens following an accident, but it might occur without an accident.

In this book, hazard management planning is defined narrowly. It covers only the planning aspects of chemical hazard management. The broader definition of hazard management planning would include, in addition to chemical hazard management, natural hazards such as flood, forest fire, drought, and earthquake. Regardless of the focus, however, the planning concept for the two categories of hazards (i.e., chemical and natural) is basically the same.

This book considers the scope of hazard management planning as extending beyond the hazardous waste domain to include all chemicals under the hazardous material definition. *Hazardous material* is defined as any substance or material in quantities and form that may pose an unreasonable risk to health and safety when transported in commerce. *Toxic chemicals* under the Superfund Amendments and Reauthorization Act (SARA) Title III, *hazardous substance* under the Comprehensive Environmental Response, Compensation, and Liability Act (CERCLA), and *hazardous waste* under the Resource Conservation and Recovery Act (RCRA) Subtitle C are regulated as subgroups of hazardous material.

Although the term *hazard management planning* — either in the context of a chemical hazard or a natural hazard — has a different connotation than the terms *emergency planning* or *disaster management planning,* they are often used synonymously. Godschalk and Brower (1985)[198] assert that *hazard management* deals with **long-term** hazard reduction goals. Hazard management focuses on improving community capabilities to withstand future disasters. This requires effective strategic skills in planning, policy design, and implementation. *Disaster* or *emergency management* is focused on improving disaster operations by providing an emergency coordination and decision system to cope with disaster-caused disruptions. These **short-term** operations require effective tactical skills in communication, command, and control.

In the Webster's Collegiate Dictionary, the term **hazard** is defined as a source of danger. The conversion of that source into an actual occurrence of loss, injury, or some form of damage is called an **event.** If an event represents a circumstance that calls for immediate action, the event is known as an **emergency**; but if it brings great damage, loss, or destruction, the event is called a **disaster.** The definition implies that a disaster is an emergency event, but not all emergencies are disasters. Two distinct features that make chemical or hazardous waste emergencies and disasters unique from a safety standpoint are that the probability of occurrence is low, but the consequences could be severe if emergencies or disasters happen (Charles and Kim, 1988[199]; Waller and Covello, 1984[200]).

Kasperson and Pijawka (1985)[201] define hazard management as a purposeful activity by which a firm or an agency informs itself about hazards, decides what to do about them, and implements measures to control them or to mitigate their consequences. In terms of management processes, the definition refers to the following four activities: prevention, preparedness, response, and recovery (Petak, 1985[202]; Zimmerman, 1985[203]).

Prevention is defined as any activity designed to reduce or prevent accidents and emergencies from occurring. Examples are:

- developing and updating an emergency plan,
- designing and monitoring a facility operating procedure, and
- complying with regulatory requirements.

Preparedness refers to a set of activities to test the effectiveness of a prevention program. These activities include:

- personnel training,
- monitoring and having emergency resources and interagency coordination, and
- conducting disaster drills.

Response involves life-saving and protection activities that are conducted during an emergency. Examples are:

- firefighting,
- spill control,
- warning notification and communication,
- facility shutdown, and
- evacuation.

Recovery embodies all activities necessary to bring the organization back to normal operations, such as:

- cleanup and restoration,
- incident investigation, and
- damage assessment.

Often, the prevention and preparedness steps can last years before an actual emergency occurs. If an emergency were to occur, the response phase would lead to the recovery phase — which would be followed by an entirely new hazard management cycle, beginning with prevention. Kelly (1989)[204] argues that these four phases are interrelated and cyclical in nature; a full cycle may take decades to complete.

The final episodes of the cycle, the response and recovery capabilities, depend very much on the quality of the prevention and preparedness activities. This means that developing and maintaining an emergency response plan represents a critical activity for the success of emergency response.

This chapter focuses on the planning activities of the above four hazard management processes. We feel it is necessary that, prior to discussing the structure and content of an emergency response and hazard management plan, we have familiarity with the regulatory requirements of the plan.

REGULATORY BACKGROUND

There are a number of federal laws that require the development of emergency plans for businesses. The applicability of those requirements depends on the type and nature of the business. Most businesses are subject to the requirements of the **Occupational Safety and Health Administration** (OSHA), RCRA Subtitle C, and SARA Title III.

Several states are also requiring companies to develop emergency plans. Notable among these state laws are the Illinois Chemical Safety Act and the New Jersey Toxic Catastrophe Prevention Act (Kelly, 1989).[205] The Illinois law requires industrial companies that deal with hazardous materials to develop and submit emergency response plans to their local disaster service agencies. Employee training represents a major element of the plan. The New Jersey law is focused on the prevention of catastrophic accidents, and therefore, its requirements revolve around prevention activities such as risk assessment and process control. The law still requires facilities to have comprehensive risk management programs that include detailed emergency response procedures and plans.

The federal OSHA requires businesses with more than ten employees to develop and maintain employee emergency and fire prevention plans. Title 29 of the Code of Federal Regulations (29 CFR) Section 1910.38 outlines the requirements for the plan, and 29 CFR 1910.120 describes standards for employees engaged in hazardous material operations — including training for both on-site and off-site emergency responders. 29 CFR 1910.120 contains the **Hazardous Waste Operations and Emergency Response** (HAZWOPER) OSHA standard, which requires health and safety programs for treatment, storage, and disposal (TSD) facility workers, hazardous waste site workers, emergency response workers, and underground storage tank (UST) workers.

Under OSHA, manufacturing and non-manufacturing industries are required to prepare and maintain Material Safety Data Sheets (MSDS) that contain concise information specific to a single chemical or chemical mixture (29 CFR 1910.1200).

There are over 500,000 chemicals identified under OSHA (U.S. EPA, 1991).[206] All facilities are required to submit MSDS information regardless of the amount of the chemicals involved. 29 CFR 1910.1200 is OSHA's **Hazard Communication Standard**, which establishes uniform requirements to ensure that the hazards of chemicals handled in the workplaces are evaluated and that this hazard information is transmitted to affected employers and exposed employees. The transmission of information is achieved through labeling, MSDS reporting, and training.

The MSDS is designed to communicate to the users of hazardous chemicals precisely what health and physical hazards are associated with these substances in the workplace. For each chemical listed in the MSDS, applicable hazard categories must be included. OSHA established 23 hazard categories. Based on this categorization, the EPA developed five different grouping schemes. The EPA's **hazard categories** are fire hazard, reactive hazard, sudden release of pressure hazard, immediate (acute) health hazard, and delayed (chronic) health hazard. Table 9.1 compares the EPA's hazard categories with those categories established under OSHA. Emergency responders may find the MSDS to be an invaluable source of information in responding to a fire, explosion, spill, or leak involving hazardous chemicals.

RCRA Subtitle C requires large-quantity generators (LQGs) and companies that own or operate hazardous waste TSD facilities to prepare contingency plans to mitigate health and environmental damage from fires, explosions, or any unplanned sudden or non-sudden release of hazardous waste or hazardous constituents to the air, soil, or water. In a less exhaustive procedure, specified in 40 CFR 262.34(d), small-quantity generators (SQGs) are also required to plan for emergency situations.

The RCRA contingency plan requirements are found in 40 CFR Part 264 Subpart D and Part 265 Subpart D. The subparts describe the content of the contingency plan, the availability and accessibility of the plan, the plan review and amendment requirements, the emergency coordinator designation, and emergency procedures.

Specific provisions on prevention activities are found in 40 CFR Part 264 Subpart C and Part 265 Subpart C. The subparts describe the requirements related to the design and operation of the facility, the required equipment, and the procedures for the testing and maintenance of equipment. The subparts also mandate the facility to make arrangements with local authorities to familiarize police, fire departments, and emergency response teams with the facility, the hazardous chemicals handled, and other emergency-related information.

SARA Title III, also known as the **Emergency Planning and Community Right-to-Know Act** (EPCRA) **of 1986** is intended to bolster emergency planning efforts at the state and local levels, and to provide citizens and governmental agencies with information concerning potential chemical hazards in their communities through "community right-to-know" reporting mechanisms. As the name suggests, EPCRA consists of two distinct components: emergency planning and community right-to-know.

Table 9.1 Hazard Category Comparison for Reporting under Sections 311 and 312 SARA Title III

EPA's hazard categories	OSHA's hazard categories
Fire hazard	Flammable
	Combustible liquid
	Pyrophoric
	Oxidizer
Sudden release of pressure	Explosive
	Compressed gas
Reactive	Unstable reactive
	Organic peroxide
	Water reactive
Immediate (acute) health hazard	Highly toxic
	Toxic
	Irritant
	Sensitizer
	Corrosive
	Other hazardous chemicals with an adverse effect with short-term exposure
Delayed (chronic) health hazard	Carcinogens
	Other hazardous chemicals with an adverse effect with long-term exposure

The *emergency planning component (§301-§303) of EPCRA* is designed to develop state and local governments' emergency preparedness and response capabilities through better coordination and planning. The act requires the governor of each state to designate a State Emergency Response Commission (SERC). The SERC will designate local emergency districts and will appoint Local Emergency Planning Committees (LEPCs) for each district. Louisiana was the first state to establish a Commission under EPCRA (Environmental Management Services, 1987).[207] As of 1990, all governors have established SERCs and have designated over 4000 local emergency districts. Thirty-five states chose counties as the basic district designations, ten states designated sub-state planning districts, and five SERCs designated the entire state as a district (U.S. EPA, 1990).[208]

To assist the LEPC in preparing and reviewing plans, the National Response Team (NRT) published a guidance document entitled *Hazardous Materials Emergency Planning Guide* in March 1987. The NRT is composed of 14 federal agencies with emergency response responsibilities. At the state level, a federally established Regional Response Team (RRT), which is composed of state representatives and federal regional officials, provides assistance to the LEPC upon request.

There are provisions in 40 CFR that mandate the owner or operator of a facility to notify the LEPC and SERC likely to be affected if there is a release into the environment of a listed extremely hazardous substance that exceeds the reportable quantity for that substance. The term *facility* in this provision refers to buildings, structures, equipment, and other stationary items at which there is a certain significant amount of hazardous substance as specified in the provision. For purposes of emergency release notification, the term *facility* includes motor vehicles, rolling stock, and aircraft.

The list of extremely hazardous substances and the respective reportable quantities is found in 40 CFR Part 355. Approximately 360 chemicals are classified as extremely hazardous substances. Any facility that has any extremely hazardous substance equal to or in excess of the amount of its threshold planning quantity shall provide notification to the SERC that it is a facility subject to the EPCRA emergency planning requirements (40 CFR 355.30[b]). The threshold planning quantity of each extremely hazardous substance is provided in 40 CFR Part 355.

In addition to mandatory notification, facilities are also subject to three other community right-to-know reporting requirements. These requirements are specified in the *Community Right-to-Know component of EPCRA*. The first requirement (§311 of EPCRA) is detailed in 40 CFR 370.21, which requires facilities that must prepare MSDS under OSHA regulations to submit either copies of their MSDS or a list of MSDS chemicals to the LEPC and SERC. For each chemical listed in the MSDS, the related hazard categories must be included.

The second requirement (§312 of EPCRA), specified in 40 CFR 370.20, requires facilities to submit aggregate information for each hazard category in the "Tier I Form." The form must be submitted annually to the LEPC, SERC, and fire department with jurisdiction. Tier I information is an estimate of the amount of chemicals (maximum amount and daily average) for each hazard category present at the facility during the preceding calendar year, and the general location of hazardous chemicals in each hazard category. The "Tier II Form," which is found in 40 CFR 370.25, may be sent by the facility instead of a Tier I Form. (A facility may either submit a Tier I Form or a Tier II Form.) The Tier II Form is an expansion of the Tier I Form and includes the chemical names of each hazard category. Because of the inclusion of the chemical names, many states have mandated the use of the Tier II Form by the regulated community.

The third reporting provision (§313 of EPCRA) is known as the Toxic Release Inventory (TRI) requirement. Facilities subject to TRI reporting are required to complete a TRI Form, also known as "Form R," for specified chemicals. The form must be submitted to the EPA, the SERC, and the designated state agency on or before July 1 of each year (40 CFR 372.30[d]). The designated state agency may be the state environmental agency. These reports should reflect releases of toxic chemicals into the air, water, or land during the preceding calendar year.

The TRI reporting requirements apply to owners or operators of manufacturing facilities that have ten or more full-time employees and that manufacture, import, process, or otherwise use a listed toxic chemical in excess of an applicable threshold quantity of that chemical. In the early 1990s, there were *322 toxic chemicals* subject to TRI reporting, i.e., 302 individual toxic chemicals and 20 categories of chemical compounds (U.S. EPA, 1990).[209] The EPA's most recent proposals will bring this list to over 600 chemicals. The current threshold quantity for a chemical manufactured or processed in a facility is 25,000 lb in a year; the threshold quantity for a chemical used in a facility is 10,000 lb in a year. By an Executive Order on August 3, 1993, the TRI reporting requirements also apply to federal facilities (Federal Register, 1993).[210]

Form R requests facility information, chemical use, chemical releases, off-site transport, waste treatment, and waste minimization information. The form must be completed for each TRI chemical the facility manufactured or used. Based on the data submitted by regulated facilities, the EPA establishes and maintains a national toxic chemical inventory. The public is able to access this national data base and obtain the information through printed document, computer software, and other means of data transmittal.

As far as planning is concerned, the legal provisions that require emergency response and hazard management planning for hazardous waste handlers are found in RCRA Subtitle C. The legal provision mandating such planning for the public sector is SARA Title III. Please note that under the Clean Air Act Amendments of 1990, the EPA is expanding regulations that will require development and implementation of risk management planning at facilities that manufacture, process, use, store, or otherwise handle regulated substances in quantities that exceed specified thresholds (Federal Register, 1993).[211] The list of regulated substances and threshold quantities were established in 40 CFR Parts 9 and 68 in 1993. In the next section, we will discuss the structure and content of contingency planning as required by RCRA.

DEVELOPING A FACILITY CONTINGENCY PLAN

Accidents can and do happen; and when they occur, our instinct is to protect ourselves and our property. In order to effectively fulfill this responsibility, it is necessary to be prepared in advance for such events. What we need is some sort of emergency or contingency plan.

The contingency plan describes the actions that facility personnel must take to minimize hazards to human health or the environment from fires, explosions, or any unplanned release of hazardous waste or hazardous constituents to the air, soil, or water (40 CFR 264.52[a] and 265.52[a]). The provisions of the contingency plan must be carried out immediately whenever there is a fire, explosion, or release which could threaten human health or the environment.

As a preventive measure, any facility is required to be designed, constructed, maintained, and operated to minimize the possibility of a fire, explosion, or release (40 CFR 264.31 and 265.31). A facility must also have aisle space to allow the unobstructed movement of personnel, fire protection equipment, spill control equipment, and decontamination equipment to any area of facility operation in an emergency (40 CFR 264.35 and 265.35). In addition, whenever hazardous waste is being poured, mixed, spread, or otherwise handled, all personnel involved in the operation must have immediate access to an internal alarm or emergency communication device (40 CFR 264.34 and 265.34).

RCRA regulations (40 CFR 264 Subparts C and D, and 265 Subparts C and D) provide guidance on how to prepare a facility contingency plan by providing a list of factors that should be included in the plan. The required components are (1) coordination agreements, (2) on-site emergency equipment, (3) emergency coordinators, (4) emergency response procedures, and (5) an evacuation plan.

Coordination Agreements

As one of the requirements, the contingency plan must describe arrangements agreed upon by local police departments, fire departments, hospitals, emergency response contractors, and state/local emergency response teams to coordinate emergency services (40 CFR 264.52[c] and 265.52[c]). Where more than one police or fire department might respond to an emergency, agreements designating primary emergency authority to a specific police department and a specific fire department must be made.

The arrangements are necessary to familiarize the responding agencies and teams with the layout of the facility, the properties and associated hazards of hazardous waste handled at the facility, the types of injuries or illnesses that could result from fires, explosions, or releases at the facility, places where facility personnel would normally be working, entrances to and roads inside the facility, and possible evacuation routes. Where state/local authorities decline to enter into such arrangements, the facility must document the refusal in the operating record (40 CFR 264.37 and 265.37).

On-Site Emergency Equipment

The contingency plan must include a list of all emergency equipment available at the facility, their locations, a physical description of each item on the list, and a brief outline of its capabilities (40 CFR 264.52[e] and 265.52[e]). The list must be kept up-to-date. The required on-site emergency equipment are (40 CFR 264.32 and 265.32):

1. An alarm system capable of providing immediate emergency instructions to facility personnel in the form of voice or signal.
2. An internal communications system capable of summoning emergency assistance from local police departments, fire departments, or state/local emergency response teams. Examples of such devices are a telephone, an intercom, or a hand-held two-way radio.
3. Portable fire extinguishers, fire control equipment (including special extinguishing equipment such as that using foam, inert gas, or dry chemicals), spill control equipment (e.g., absorbent pillows), and decontamination equipment. The list may also include self-contained breathing apparatus (SCBA) and chemical detection equipment for

monitoring releases of hazardous substances, such as combustible gas analyzers or oxygen meters.

4. Water at adequate volume and pressure to supply water-hose streams, automatic sprinklers, water spray systems, or foam-producing equipment.

All facility communications, alarm systems, fire protection equipment, spill control equipment, and decontamination equipment must be tested and maintained as necessary to assume its proper operation in time of emergency (40 CFR 264.33 and 265.33).

Emergency Coordinators

A list of names, addresses, and office and home telephone numbers of all persons qualified to act as the emergency coordinator must be provided in a contingency plan (40 CFR 264.52[d] and 265.52[d]). The list must be kept up-to-date.

The emergency coordinator must be thoroughly familiar with all aspects of the facility's contingency plan, all operations and activities at the facility, the location and characteristics of waste handled, the location of all records within the facility, and the facility layout. In addition, this person must have the authority to commit the resources needed to carry out the contingency plan (40 CFR 264.55 and 265.55). Responsibilities for the emergency coordinator vary from one company to another, depending on factors such as the type and complexity of the facility and the type and variety of wastes handled by the facility.

Facility emergency responders must be trained in recognizing health and safety hazards to a level of competence that allows them to protect themselves and other employees. Emergency response training must be completed before the employee is called upon to perform in a real emergency. The training must be rehearsed regularly and documented for each employee. Examples of the types of training are:

- elements of the facility contingency plan,
- standard operating procedures specifically written for the employee's position,
- methods used to minimize the risk from chemical hazards,
- personal protective equipment and safe use of control equipment,
- procedures for handling emergency situations and appropriate responses to injury or overexposure from chemical hazards to themselves and other employees,
- recognition of symptoms from overexposure, and
- techniques of coordination with other employees.

Where more than one person is listed, one must be named as primary emergency coordinator, and the others must be listed in the order in which they will assume responsibility as alternates. Normally, the emergency coordinator (or alternate) has the responsibility to activate internal alarms.

Emergency Response Procedures

The emergency response procedures are specified in 40 CFR 264.56 and 265.56. The following paragraphs outline the procedures:

- Whenever there is an imminent or actual emergency situation, the emergency coordinator (or designee) must immediately activate the internal facility alarm or communication system to notify all facility personnel, and notify appropriate state/local agencies with designated response roles if their help is needed.
- Whenever there is a release, fire, or explosion, the emergency coordinator must immediately identify the character, exact source, amount, and a real extent of any released materials through observation or review of facility records or manifests and, if necessary, by chemical analysis.
- Concurrently, the emergency coordinator must assess possible hazards to human health or the environment that may result from the release, fire, or explosion. In the assessment, the coordinator must consider direct and indirect effects. The identification of transport routes and shipment methods will assist in the overall hazard analysis. The provision of a legible site map (8"× 11") of the facility showing locations of buildings, general areas of storage, chemical and hazardous waste storage locations, and roadway entrances (including street names) is also helpful.
- If the emergency coordinator determines that the facility has had a release, fire, or explosion which could threaten human health or the environment outside the facility, the coordinator must do the following:

    — Immediately notify either the government official designated as the on-scene coordinator for that geographical area, or the National Response Center using their 24-hour toll-free number (800) 424-8802.
    — Notify appropriate local authorities if evacuation of local areas may be advisable.

- During an emergency, the coordinator must take all reasonable measures necessary to ensure that fires, explosions, and releases do not occur, recur, or spread to other hazardous materials at the facility.
- If the facility stops operations in response to a fire, explosion, or release, the coordinator must monitor for leaks, pressure buildup, gas generation, or ruptures in valves, pipes, or other equipment.
- Immediately after an emergency, the emergency coordinator must provide for treating, storing, or disposing of recovered wastes, contaminated soil or surface water, or any other material that resulted from the release, fire, or explosion at the facility.

- The coordinator must ensure that, until cleanup procedures are completed, no waste that may be incompatible with the released material is treated, stored, or disposed of in the affected areas of the facility. Also, all emergency equipment must be cleaned and ready for its intended use before operations are resumed.
- Within 15 days after the incident, the emergency coordinator must submit a written report on the incident to the EPA regional office.

Evacuation Plan

The contingency plan must include an evacuation plan for facility personnel where there is a possibility that evacuation could be necessary (40 CFR 264.52[f] and 265.52[f]). The evacuation plan describes the signal(s) to be used to begin evacuation, the designated evacuation routes, and the alternative routes.

The emergency coordinator is responsible for determining which emergency situations require plant evacuation. If area evacuation is initiated by the coordinator, the follow-up actions must be elaborated in the plan. They include, for example, the designation of rally points, arrangement of vehicular traffic, plant reentry, etc.

The contingency plan must be reviewed and amended whenever applicable regulations are revised; whenever the facility changes its design, construction, operation, or maintenance; or whenever other circumstances occur that either increase the potential for fires, explosions, or releases, or change the response necessary in an emergency (40 CFR 264.54 and 265.54). When a plan fails in an emergency, it also is required to be amended.

A plan failure may occur when the response time and the response actions are inadequate. Avoiding this type of mistake requires the facility to update the plan periodically and have it field-tested. Briese and Griffith (1991)[212] state that a contingency plan that was written but never tested is useless, and Baldwin (1992)[213] cautions that having a written contingency plan may give everyone in the company a false sense of security. Therefore, Cahill and Kane (1992)[214] encourage that emergency response training and drills be undertaken seriously and routinely. They underline the critical nature of the training and drills because of the fact that the first half-hour of a release represents a time of response action, not deliberation.

As part of hazard management planning protocol, federal regulations require that a copy of the contingency plan and all amendments to the plan be maintained at the facility (40 CFR 264.53 and 265.53). The plan must also be submitted to all local police departments, fire departments, hospitals, and state/local emergency response teams that may be called upon to provide emergency services.

COMMUNITY EMERGENCY RESPONSE PLAN

The EPCRA makes it mandatory for the LEPC to prepare an emergency plan for possible releases of hazardous substances and for facilities to cooperate in this planning process. The intent of this requirement is to develop emergency preparedness and response capabilities through better coordination and planning. A graphic representation of the flow of information required by EPCRA is shown in Figure 9.1.

The EPCRA specifies that the SERC is responsible for the following activities:

1. supervising and coordinating activities of the LEPCs,
2. establishing procedures for receiving and processing public requests for information collected under the Act, and
3. reviewing community emergency plans.

The LEPC's primary responsibility is to develop and update the community emergency response plan by considering the resources available for preparing for and responding to a potential chemical accident. To assist the LEPC in preparing and reviewing plans, Congress required the National Response Team (NRT), which is composed of 14 federal agencies with emergency preparedness and response responsibilities, to publish guidance on emergency planning. Guidance was published in 1987[215] by the NRT to fulfill this mandate.

Upon completion of the community emergency plan and its annual review by the LEPC, the plan must be reviewed by the SERC. If requested by the LEPC, the Regional Response Team (RRT), which is composed of federal regional officials and state representatives, may review the plan and provide assistance.

The 1987 NRT guidance document specifies that community emergency plans must include the following components:

1. identification of facilities that handle chemicals, and transportation routes of extremely hazardous substances;
2. emergency response procedures, both on-site and off-site;
3. designation of a community coordinator and facility coordinator(s) to implement the plan;
4. emergency notification procedures;
5. methods for determining the occurrence of a release and the probable affected area and population;
6. description of community and industry emergency equipment and facilities, and the identity of persons responsible for them;
7. evacuation plans;
8. description and schedules of a training program for emergency response to chemical emergencies; and
9. methods and schedules for exercising emergency response plans.

In terms of the planning process, community emergency planning consists of five major tasks:

1. To search out and *review all existing emergency plans.* The sources for this task include state/local emergency management agencies, fire

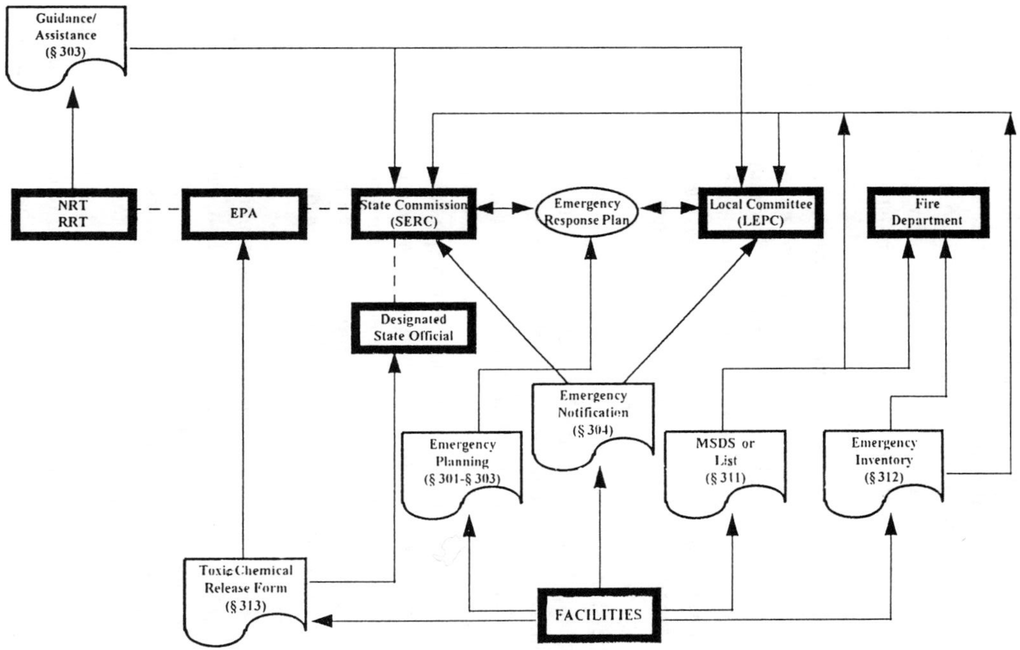

Figure 9.1 Major Information Flow/Requirements. (From National Response Team, *Hazardous Materials Emergency Planning Guide*, NRT, Washington, D.C., 1987.)

departments, police departments, transportation agencies, environmental agencies, public health agencies, public service agencies, volunteer groups, associations, and regional offices of federal agencies.

2. To *conduct hazard analysis for setting priorities for planning.* This task may include hazard identifications, vulnerability analysis, and risk analysis, or simply the identification of the nature and location of hazards in the community. Vulnerability is the susceptibility of life, property, and the environment to adverse effects (injury or damage) if a hazard manifests its potential. Risk is defined as the probability that the adverse effect will occur. Developing a complete hazard analysis may not be desirable; therefore, planners must consider the level of thoroughness that is appropriate. They may ask local facilities whether they have already completed a facility hazards analysis.
3. To *assess preparedness, existing prevention measures, and response capabilities* of the facility, transporter, and community. The facility assessment should provide information on the status of the facility's contingency plan and whether the plan is consistent with any community emergency plan, the scope of the emergency training plan, and the status of the emergency response equipment. Information required from the transporter assessment includes cargo data, type of equipment, cleanup capabilities, and the level of safety training for operators. The community assessment should provide information about which local agencies are involved in the emergency response network, the capacity and level of expertise of medical service, mutual aid agreements if any, community points of contact and their responsibilities, the status of planning and coordination for emergency preparedness, the types and availability of emergency response equipment and materials, specialized emergency response teams if any, existing procedures for protecting citizens during emergencies, communication systems, and other similar pertinent facts.
4. To *complete the hazardous materials plan,* which describes the personnel, equipment, and procedures to be used in case of an unplanned release of a hazardous material — if one does not already exist. Many local jurisdictions already have emergency plans for various types of hazards (e.g., floods, earthquakes, and hurricanes, as well as hazardous materials incidents); however, if a community does not have the resources, time, or capability readily available to undertake multi-hazard planning, it may wish to produce a single-hazard plan addressing hazardous materials. Table 9.2 provides a sample outline of a hazardous materials emergency plan. The NRT cautions that the sample is not a model. It is not meant to constrain any community.
5. To *develop an ongoing program for implementation, maintenance, training, and practice.* This means that any emergency plan must be monitored, evaluated, and kept up-to-date through the review of actual responses, simulation practices, and regular collection of new data. Updated

Table 9.2 Sample Outline of a Hazardous Materials Emergency Plan

A. Introduction
  1. Incident Information Summary
  2. Promulgation Document
  3. Legal Authority and Responsibility for Responding
  4. Table of Contents
  5. Abbreviations and Definitions
  6. Assumptions/Planning Factors
  7. Concept of Operations
     a. Governing Principles
     b. Organizational Roles and Responsibilities
     c. Relationship to Other Plans
  8. Instructions on Plan Use
     a. Purpose
     b. Plan Distribution
  9. Record of Amendments

B. Emergency Assistance Telephone Roster

C. Response Functions
  1. Initial Notification of Response Agencies
  2. Direction and Control
  3. Communications (among Responders)
  4. Warning Systems and Emergency Public Notification
  5. Public Information/Community Relations
  6. Resource Management
  7. Health and Medical Services
  8. Response Personnel Safety
  9. Personal Protection of Citizens
     a. Indoor Protection
     b. Evacuation Procedures
     c. Other Public Protection Strategies
  10. Fire and Rescue
  11. Law Enforcement
  12. Ongoing Incident Assessment
  13. Human Services
  14. Public Works
  15. Others

D. Containment and Cleanup
  1. Techniques for Spill Containment and Cleanup
  2. Resources for Cleanup and Disposal

E. Documentation and Investigative Follow-Up

F. Procedures for Testing and Updating Plan
  1. Testing the Plan
  2. Updating the Plan

Table 9.2 Sample Outline of a Hazardous Materials Emergency Plan (Continued)

G. Hazards Analysis (Summary)
H. References
    1. Laboratory, Consultant, and Other Technical Support Resources
    2. Technical Library

From National Response Team, *Hazardous Materials Emergency Planning Guide,* NRT, Washington, D.C., 1987.

plans should reflect any recent changes in the economy, land use, permit waivers, technology, response capabilities, hazardous materials present, laws and regulations, road configurations, population, and facility location. In addition to the periodic updates, incident reviews, practice sessions, and training are vital elements to ensure effective planning.

HAZARDS ANALYSIS AND COMMUNITY VULNERABILITY

A hazard analysis is a critical component of planning for hazardous materials releases. Because of its significance, the last section of this chapter is devoted to this particular topic. Conceptually, hazard analysis consists of three major activities:

1. Hazard identification.
2. Vulnerability analysis.
3. Risk analysis.

The *hazard identification* provides information about the facility and transportation situations that have the potential for causing injury to life or causing damage to property and the environment due to a hazardous materials spill or release. It must indicate the types and quantities of hazardous materials located in the facility or transported through a community, the location of the facilities and the shipment origin, destination, and routes, and the nature of hazard most likely to accompany hazardous materials releases, such as fire or explosions. The hazard identification should result in a compilation of those situations that may pose the most serious threat of damage to the community. Location maps, charts, and other visual displays are excellent means of presenting this information.

The *vulnerability analysis* identifies zones in the community that are susceptible to damage should a hazardous materials release occur. The analysis should provide information about the characteristics of the vulnerable zone and the conditions that influence vulnerability, such as the size of release, topography, and wind direction. The characteristics of the vulnerable zone can be indicated by the size and types of land use (residences, employment centers, hospitals, schools, and sensitive natural areas), including essential support systems (water supply, power, and food).

The *risk analysis* examines the probability of a hazardous materials release and the range of possible consequences, measured by such indicators as the number of fatalities and injuries, or the monetary losses due to property and environmental damage that are taking place in any vulnerable zone in the community. Specifically, a risk analysis enables planners to obtain information on:

1. the probability that a release will occur by type of hazardous material;
2. the harm to people (fatalities or acute, delayed, and chronic illnesses);
3. the damage to property (permanent, temporary, or repairable); and
4. the damage to the environment (permanent or recoverable).

Performing a hazard analysis can require specialized expertise. The NRT advises that communities should not be overly concerned with developing elaborate quantitative release probabilities; instead, occurrence probabilities can be simply described in relative terms, such as low, moderate, and high. The emphasis should be on developing reasonable estimates based on the best available expertise.

Table 9.3 provides an example of a very simple hazards analysis for a hypothetical community. In this table, hazard A (the use of chlorine in a water treatment plant), hazard B (transportation of ammonia), and hazard C (the manufacture of liquid MIC) are identified as three major hazards in the community. Once the hazard analysis is completed, it represents an essential tool in the planning process, because it assists planners in determining the level of detail that is necessary, the types of response to emphasize, and the priority hazards or areas for planning.

Risk itself, however, does not entirely represent community vulnerability. Referring to community vulnerability as the safety status of a community, Gabor and Griffith (1980)[216] define community vulnerability as a function of community risk and community preparedness. Utilizing this concept, Pijawka et al. (1984)[217] introduce a typology of four categories of community vulnerability (Figure 9.2).

Cell I in Figure 9.2 depicts communities characterized by a low community risk with a relatively large investment in emergency preparedness. This represents communities that have overestimated risk. In such cases, excessively high costs of preparedness may not be a cost-effective strategy.

Cell II represents high-risk and high-preparedness communities. Line AB in Cell II reflects a relationship between community risk and level of preparedness; i.e., the larger the risks faced by a community, the more investment one would want to place in mitigation planning to reduce the consequences of risk. However, the relationship between costs of preparedness and the reduction of vulnerability may not be linear because of the high initial cost of preparedness, such as the first purchase of equipment or the preparation of the first emergency response plan (Pijawka et al., 1985).[218]

Cell III illustrates a community with a high risk level but a low degree of preparedness. Among the four categories, this is the most vulnerable community. Cell IV indicates a community with low preparedness but also a low risk level.

Table 9.3 Example Hazards Analysis for a Hypothetical Community

	Hazard A	Hazard B	Hazard C
1. Hazard identification (major hazards)			
a. Chemical	Chlorine	Ammonia	Liquid methyl isocyanate (MIC)
b. Location	Water treatment plant	Tank truck on local interstate highway	Pesticide manufacturing plant in nearby semi-rural area
c. Quantity	2000 lb	5000 lb	5000 lb
d. Properties	Poisonous; may be fatal if inhaled. Respiratory conditions aggravated by exposure. Contact may cause burns to skin and eyes. Corrosive. Effects may be delayed.	Poisonous; may be fatal if inhaled. Vapors cause irritation of eyes and respiratory tract. Liquid will burn skin and eyes. Contact with liquid may cause frostbite. Effects may be delayed. Will burn within certain vapor concentration limits and increase fire hazard in the presence of oil or other combustible materials.	Causes death by respiratory distress after inhalation. Other health effects would include permanent eye damage, respiratory distress, and disorientation, Explosive. Extremely flammable.
2. Vulnerability analysis			
a. Vulnerable zone	A spill of 2000 lb of chlorine from a storage tank could result in an area of radius 1650 ft (0.3 mi) where chlorine gas may exceed the level of concern.	A spill of 5000 lb of ammonia resulting from a collision of a tank truck could result in an area of radius 1320 ft (0.25 mi) where ammonia exceeds its level of concern.	A spill of 5000 lb of methyl isocyanate could affect an area of radius 3300 feet (0.6 miles) with MIC vapors exceeding the level of concern (assuming that the liquid is hot when spilled, the tank is not diked, and the MIC is at 100% concentration).
b. Population within vulnerable zone	Approximately 500 residents of a nursing home; workers at small factory.	Up to 700 persons in residences, commercial establishments, or vehicles near highway interchange. Seasonal influx of visitors to forest preserve in the fall.	Up to 200 workers at the plant and 1000 children in a school.

Table 9.3 Example Hazards Analysis for a Hypothetical Community (Continued)

	Hazard A	Hazard B	Hazard C
c. Private and public property that may be damaged	Facility equipment, vehicles, and structures susceptible to damage from corrosive fumes. Community's water supply may be temporarily affected given that the facility is its primary supplier. Mixture with fuels may cause an explosion.	25 residences, 2 fast food restaurants, one 30-room motel, a truck stop, a gas station, and a mini-market. Highway and nearby vehicles may be susceptible to damage from a fire or explosion resulting from the collision.	Runoff to a sewer may cause an explosion hazard because MIC reacts violently with water.
d. Environment that may be affected	Terrestrial life.	Adjacent forest preserve is highly susceptible to forest fires especially during drought conditions.	Nearby farm animals.
3. Risk analysis			
a. Probability of hazard occurrence	Low — Chlorine is stored in an area with leak detection equipment in 24-hour service with alarms. Protective equipment is kept outside storage room.	High — Highway interchange has a history of accidents due to poor visibility of exits and entrances.	Low — Facility has up-to-date containment facilities with leak detection equipment, and an emergency plan for its employees. There are good security arrangements that would deter tampering or accidents resulting from civil uprisings.
b. Consequences if people are exposed	High levels of chlorine gas in the nursing home and factory could cause death and respiratory distress. Bedridden nursing home patients are especially susceptible.	Release of vapors and subsequent fire may cause traffic accidents. Injured and trapped motorists are subject to lethal vapors and possible incineration. Windblown vapors can cause respiratory distress for nearby residents and business patrons.	If accident occurs while school is in session, children could be killed, blinded, and/or suffer chronic debilitating respiratory problems. Plant workers would be subject to similar effects at any time.
c. Consequences for property	Possible superficial damage to facility equipment and structures from corrosive fumes (repairable).	Repairable damage to highway. Potential destruction of nearby vehicles due to fire or explosions.	Vapors may explode in a confined space causing property damage (repairable). Damage could result from fires (repairable).

Table 9.3 Example Hazards Analysis for a Hypothetical Community (Continued)

	Hazard A	Hazard B	Hazard C
d. Consequences of environmental exposure	Possible destruction of surrounding fauna and flora.	Potential for fire damage to adjacent forest preserve due to combustible material (recoverable in the long term).	Farm animals and other fauna could be killed or suffer health affects necessitating their destruction or indirectly causing death.
e. Probability of simultaneous emergencies	Low	High	Low
f. Unusual environmental conditions	None	Hilly terrain prone to mists, thus creating adverse driving conditions.	Located in a 500-year river flood plan.

From National Response Team, *Hazardous Materials Emergency Planning Guide,* NRT, Washington, D.C., 1987.

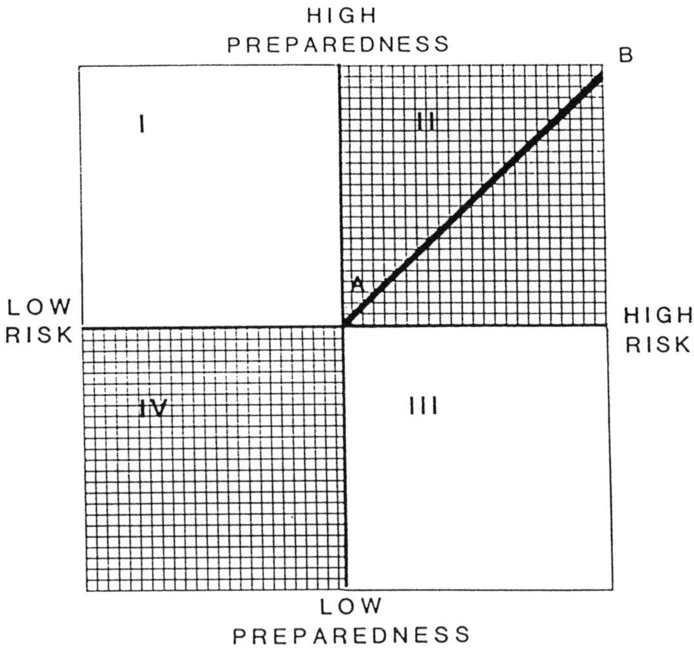

Figure 9.2 Risk-Preparedness Scales for Communities. (From Pijawka, K. David, Foote, Steve, and Soesilo J. Andy, *Risk Assessment of Transporting Hazardous Materials, A Conceptual Framework and Approach for Risk Management and Planning,* Arizona State University, Tempe, AZ, 1984.)

10 Citizen Participation Planning

Industrial plants are normally under the constant scrutiny of the regulatory agency and the general public. A hazardous waste accident, even of the most routine nature, will always touch off immediate public reaction. We learned from Chapter 7 that hazardous waste management facility siting efforts also stir inquiry and often result in resistance from the public.

The Resource Conservation and Recovery Act (RCRA) considers public involvement as extremely important to the success of the RCRA program. The RCRA public participation framework covers a number of major areas. Under Section 3007(b), the public is given the right to see information obtained from a facility inspection conducted by the regulatory agency. Section 3006 provides requirements that the public be given the opportunity to comment before the issuance of decisions regarding RCRA authorization to a state and the permitting of a treatment, storage, or disposal (TSD) facility. Section 7002 allows a citizen to bring a civil suit against any person or government agency — including the EPA — alleged to be in violation of any permit, standard, regulation, or provision of RCRA.

The Hazardous and Solid Waste Amendments (HSWA) of 1984 expand this citizen suit provision to include imminent and substantial endangerment actions against past and present hazardous waste handlers (Hall et al., 1991).[219] Citizen suits, however, are prohibited in the following circumstances:

- in the siting and permitting of TSD facilities,
- where the EPA is prosecuting under RCRA or Superfund authority,
- while the EPA or a state is engaged in a Superfund removal or has incurred costs to engage in a remedial action, and
- where the responsible party is conducting a removal or remedial action pursuant to an EPA order.

Citizen suits can cause very substantial damage to a company's operation. Some suits may convince the court to impose large penalties and to order permanent closure of a noncomplying RCRA facility. The above discussion is to illustrate the power that Congress has vested in citizens under RCRA provisions.

In many respects, the relationships among the business community, the government, and the general public have become less confrontational and more cooperative in recent years than a decade or two ago. While the level of trust among the three groups may not have increased, the three parties have begun to realize that working together often provides greater benefits than maintaining adversarial relationships.

Citizen participation is essentially about building and maintaining communication links with all parties involved. Citizen participation concentrates on communication. The goal of citizen participation planning is to establish effective and efficient communication among all of the stakeholders in the hazardous waste management system: business, government, and the public.

FUNDAMENTALS OF ENVIRONMENTAL COMMUNICATIONS

A company that takes the initiative in communication gives itself leverage in winning public understanding. This means going beyond compliance with environmental regulations to demonstrate full and open communication. A company's effective environmental communication can be viewed as a balance between its credibility, sound decision-making, positive actions, and community value sensitivity (Lukaszewski and Serie, 1993).[220] Of the four attributes of effective communications, developing and maintaining credibility is the core of environmental communications. Building credibility is an ongoing process. It is a commitment that all promises, direct or indirect, will be kept and exceeded wherever possible. The company can expect that this commitment may require changes in the company's plans.

Lukaszewski and Serie admit that credibility can be very hard to achieve and maintain, because this attribute represents quality expectations that are difficult to define and measure. It is a judgment formed in the hearts and minds of the public.

Credibility is a perception of how believable and how trustworthy a company is, based on the evaluation of its words and actions from the community's perspective. Harrison (1992)[221] states that if a company desires to build a company-citizen partnership, it must be based on the three fundamental components; i.e., the company must (1) open its operation for observation, (2) provide open dialogue, and (3) be open to options.

People normally distrust and fear a company on the basis of what they know. When the subject is the operation of a hazardous waste management facility and its associated risks, their perception is presumably about something very dangerous and about an uncontrollable situation. With the right planning, an **open house or plant tour**, which provides the community with an opportunity to examine the facility, can build up community acceptance and good will. Harrison argues that in some cases, these kinds of "come-look-us-over" events can help to overcome negative situations.

While the profit-making interest of a company is obvious, the company must make it clear that it also stands on the side of the public interest. The message must be concise, positive, and non-defensive, and the community must be convinced that its interests come first. The key to this, according to Harrison, is **open dialogue** with the community and an attitude of caring and honesty.

Strengthening the company-citizen partnership can also be achieved by providing the community with a list of options for solving environmental problems. The *options for a proposed project,* such as alternative locations, alternative plant designs, provisions for monitoring environmental change, and financial factors can all be put on the table for the community to discuss. While this approach may reduce the company's level of control, it still is a logical approach, because a community will usually reject the proposal that is presented in finished form. Harrison emphasizes that presentation becomes as important as the intention. The goal is to make the community feel that it has some control over its own destiny. In the final analysis, if a company is unable to gain a consensus that makes good business sense, it will ultimately have the option of abandoning its idea for the project.

As more and more information flows into the community, the key to community relations for business today is to involve the public directly in the decision-making process, and to communicate clearly, honestly, and constantly about issues that must be faced together. Harrison believes that a true partnership which is based on those three fundamental concepts is the only viable option for companies and citizens sharing the same environment in a community.

Working together in a mutually beneficial environment is the goal of a community relations program. An effective program needs hard work and dedication and demands good planning and intensive preparation. An effective program requires a strategy.

STRATEGIZING COMMUNITY RELATIONS

In strategizing a community relations program, we have to be aware that a company-citizen relationship will occur for two basic reasons: the public will become involved in the process (1) because the program appeals to their interests, or (2) because it threatens their interests (Lukaszewski and Serie, 1993).[222] Relationships that occur because of mutual interests are usually voluntary, and companies often spend far too much time on them because they are so warm and fuzzy. Relationships that grow from a perceived threat to the public's interest are usually involuntary, forced by circumstances, and must be identified early and worked on most aggressively.

This differing motivation must be considered when involving a community in a community relations program. Programs to address public concerns must be planned delicately. McDermitt (1993)[223] provides ten major steps to successful community relations. Although his strategy is written to win project approval, it can also be used for community relations.

The first element of the strategy is to **plan early**. A community relations plan crafted early in facility development can help shape public support, because nearly every facility has a good story to tell to its public, such as increased local revenue, jobs, and environmental improvements. These benefits will help win local support if the right messages are properly communicated. Failure to properly transmit the information will lead to rumors, speculation, and the spread of misinformation and opposition.

The second element is to **prepare properly**. Identify who the community residents respect the most. This action is necessary because the company will want those individuals on its side. There are a number of research tools that can be used. One of them is a public opinion research poll, which can provide information on key factors about a community, its attitudes, concerns, and credible local opinion leaders.

The **development of a comprehensive plan** is the third element. The plan clarifies the existing situation, defines the plan's objectives in clear and concise terms, analyzes barriers to implement the plan, specifies the techniques to address each barrier, outlines the activities to achieve each objective, determines resources required, and schedules plan implementation, including assignments of responsibilities of each task.

Allocating budget, i.e., dedicating sufficient finances to conduct plan implementation, represents the fourth element — followed by **being proactive**. McDermitt emphasizes that the residents of most communities are not radical and anti-development fanatics. Most have genuine concerns about their community and the impacts a facility may have on their lives. It is the responsibility of the company to demonstrate sensitivity to the community's concerns and a willingness to address them.

The sixth element of the strategy is to **package the project**. The community relations project is normally complex to the average observer. A good tool to relay the information to the community is a facility information package. Begin by describing the need for the facility in the area, avoid using the plant's technology terms and confusing statistics, and explain how the facility will protect the environment — but do not become involved in the discussion of environmental issues. The information kit may contain a question-and-answer section.

Handling the politics is the next step in the strategy. This involves identifying those officials who carry the real power in the host community and showing them how the facility benefits them and their constituencies. This is followed by **identifying supporters** who can assist the facility, for example, at public hearings. The success of a community relations program will depend largely on how effective a facility is in finding, cultivating, and mobilizing local supporters.

The ninth element is to **manage the media**. The company goal is to achieve fair and accurate coverage of the facility by the media. McDermitt emphasizes that it is not only what you say, but how you say it, that will leave a lasting impression on the public. A facility spokesperson should be professionally trained to perform this task.

The tenth element of the strategy is to **manage the hearings**. Proper planning can help the facility win public hearings. If possible, avoid any hearing at all. If not, ensure that the hearing is fair and orderly. McDermitt suggests working with the media at public hearings by providing them with the information they need to write their story, because most residents' impressions of the hearing will be what they read in the newspaper the following day.

The message that underlines the above strategy is that as the public becomes more actively involved in environmental issues that affect their lives, the business community will be required to focus more attention on building and maintaining good relationships with the public. The same message holds true for governmental agencies. With the increasing intensity of policy and societal issues, there is an apparent need for regulatory agencies to also develop and maintain good public participation programs.

DEVELOPING A PUBLIC PARTICIPATION PLAN

Whether a state is considering the implementation of a mandatory hazardous waste minimization program, a facility siting and permitting process, a corrective action, or the assessment of the need for a hazardous waste plan, public participation will play a significant role both before a program is finalized and during the implementation stage. The term *public participation* entails a broad scope of activities designed to help citizens participate in decisions, convey information, solicit citizen concerns, heighten public awareness, and motivate participation in programs (U.S. EPA, 1989).[224]

Successful public participation programs are the result of careful planning. By developing a realistic plan, decision-makers can evaluate the situation, problems, and issues, and know where best to direct their efforts and resources. A public participation plan can be defined as consisting of the following elements:

1. **Identification of issues** that need to be addressed. An effective program should be planned with the community's needs in mind. It is helpful and time- and energy-saving to examine the public participation plans of other agencies, then borrow from their successes and learn from their failures. Of similar importance is understanding the different audiences that exist within the community and determining how these diverse groups obtain information. For example, what are citizens concerned about, will they read written information, and will they respond well to public notices? Answers to these kinds of questions ensure that the appropriate messages are sent in the appropriate format.
2. **Formulation of goals** to be reached. The objectives must be concisely stated. Clearly communicate to the public how and when they will have the opportunity to participate. Include enough detail in the plan so that everyone involved in implementing the plan knows what is expected of them and when.
3. **Description of activities** to accomplish each goal. The public participation plan should be positive and should provide simple instructions on how to participate. Planned activities should satisfy the information needs of the community and should be within the budgetary and

resource limits. In some cases, it is more beneficial, for example, to conduct pilot projects for testing new ideas.

4. **Availability of resources** for each activity. Public participation planning is relatively inexpensive. However, it does require a commitment of funds and staff time to plan and coordinate a successful program.
5. Program **scheduling, coordination, and monitoring**. Opportunities for communicating with and involving the public should be initiated early in the planning process. Communication with the public and promotion of the program should be ongoing, because a public participation program is a dynamic and continuous planning effort.

There is a broad range of possible activities that can be included in a public participation plan. The activities mostly represent educational undertakings. Because of this, public participation planning is often described as public education and involvement planning. In essence, public participation implies integrating public concerns and values at *every stage* of the decision-making process. It is a dialogue — a two-way communication — that involves getting information out to the public and getting back ideas, issues, and concerns from the public.

To improve public participation planning, the EPA provides a list of information and involvement techniques for use in developing the plan (U.S. EPA, 1990).[225] **Information techniques** can be used to get information *to* the public, while **involvement techniques** are used to get information *from* the public. These two types of techniques are, respectively, discussed in the next two sections.

Information Techniques

A successful public participation program requires a well-planned public information strategy. The public needs to know the rationale of why a project or an activity is needed and the consequences if it is not implemented. They need information about alternative technology or options that are available and other facts that they can use to evaluate the project or activity appropriately.

The EPA's list of techniques for communicating information to the public include the following:

1. Provide **briefings**. The goal is to keep key elected officials informed on any new event. This can be done simply by a personal visit or a phone call.
2. Assist reporters in writing **feature stories**. Sending a news release to a newspaper is one way to get media attention, but personal contact with an editor is more likely to get someone interested.
3. Mail out **reports** directly to key individuals. When constructing a mailing list, code the names of key individuals to accommodate this mail-out. A summary of the report can be sent to the remaining names on the mailing list with a note that a copy of the full report is available upon request or at various locations for review.

4. Arrange **news conferences**. This technique stimulates the interest of the media in developing news stories. The key to a successful news conference is that the topic of the conference or the person conducting it must be newsworthy, or no one will show up.
5. Develop **newsletters**. This medium provides interested individuals with far more information than the news media. The effectiveness of a newsletter often depends on its format, i.e., attractiveness, pictures, graphics, and the use of simple, daily language.
6. Provide **newspaper inserts**. As long as the insert meets the newspaper's specifications, newspapers can deliver an insert for a moderate cost. Similar to a newsletter, an insert must be attractive and easy to read in order to attract the readers.
7. Issue **news releases**. This text should interest the media in doing a news article on the issue. The name and number of the contact person dealing with the issue must be included in a news release.
8. Arrange for **paid advertisements**. The public is normally appreciative of paid advertisements announcing a public meeting or presenting information; however, any advertisement paid with public funds, especially large ads, is also subject to criticism.
9. Make **group presentations**. A slide show should be considered, because a visual presentation is more interesting to the audience and can be done in a short time.
10. Develop **press kits**. A kit summarizes key information from a number of documents and reports. Once a press kit is finalized, deliver the kit to interested editors, and be prepared to answer questions that arise.
11. Utilize **public service announcements**. This service is provided free of charge by radio and television stations. The stations are very likely to broadcast the program and to invite the public to participate. Through this program, the audience will not only listen but also respond to the issues.

Communication is a two-way process: giving the information to and receiving the opinions of the public. Overall, the techniques described in the preceding paragraphs are designed to provide the public with specific information. Once the public has been informed, the remaining part of the communication task is to provide mechanisms by which the public can express their concerns and give opinions and suggestions. The mechanism represents a forum of public involvement.

Involvement Techniques

A public involvement process is necessary for the exchange of information. The process can be conducted in a number of ways. The EPA's list of involvement techniques for effective exchange of information includes the following:

1. Establish **task forces**. Task forces and advisory groups can serve a number of purposes, such as providing communications between key constituencies, providing a forum for reacting to the more technical aspects of a discussion, and achieving a consensus decision. It must be decided early in the process what mandate the group has in the decision-making process, i.e., whether they are purely advisory, or have some sort of authority in giving a decision.
2. Establish **focus groups** as an alternative to polls. These are small discussion groups and are used to explore people's reactions to issues. A trained moderator to lead the groups is essential to the success of using this technique.
3. Establish a **hotline**. The hotline enables people to call to ask questions as well as to comment. The key to an effective hotline is the right person receiving the calls.
4. Conduct **interviews**. Interviews are powerful because some people will provide more information in a one-on-one conversation than they will in an open forum, such as a public meeting.
5. Facilitate **workshops**. Workshops or meetings are widely used public involvement techniques. Ensure that everyone gets a chance to be heard.
6. Conduct a **plebiscite**. The ultimate test of whether a community supports a program or activity is a direct vote on it.
7. Conduct **polls**. The purpose of conducting a poll is to assess community viewpoints quantitatively. Polls, however, do not always predict outcome. They only provide a snapshot of one moment in time. They are helpful and informative but do not replace the need for some of the other involvement techniques described above.

Public Participation Pitfalls

The effectiveness of public participation does not result from using a single public involvement technique, but from combining many techniques of public involvement and public information into a unified program strategy. A few cautions, however, should be observed in developing a public participation plan. They are:

1. **Avoid public hearings**. The danger of public hearings is that they tend to exaggerate differences rather than bring people together. Workshops provide more satisfactory results.
2. **Be aware of task force limitations**. The most obvious one is the extent of its authority. Another limitation stems from the fact that advisory groups and task forces can spend more time agreeing on procedures rather than on substance.
3. **Keep public information objective**. The public information document is designed to provide the information the public needs in order for them to be involved in an informed manner. It is not aimed at selling

a particular point of view; therefore, the document must be objective, balanced, and credible.

4. **Deal straight with media**. Provide all the information objectively and factually. The media have two important characteristics for communicating information to the public. They are the most accessible sources of information, and they are seen by the public as the most credible sources of information.
5. **Provide feedback loops**. This means that if we ask the public to participate, always get back to them promptly with updated information. The primary motivation for participation is the sense that everyone in the community can have an impact. Feedback loops strengthen this motivation.
6. **Never surprise elected officials**. Keep them informed on the status of every activity first-hand.

RISK COMMUNICATION IN CITIZEN PARTICIPATION PLANNING

In general, hazardous waste management issues are considered complex because they are not only technical, but also societal, economic, and political in nature. One of the most difficult issues in citizen participation planning associated with hazardous waste management is risk communication.

The issue is difficult because in the field of environmental management, pollution is legally permitted. As long as the level of pollutant releases to the environment is below the regulated threshold limit, such releases are allowed. These allowable releases represent a risk to the public, no matter how small and insignificant the risk is. The public is actually forced to take this risk involuntarily.

In response, the public evaluates the risk in terms of their perception of the risk. For involuntary risk, they often demand a zero risk level — something that is impossible to achieve. In reality, the perceived risk is always greater than the actual risk. The EPA (1990)[226] observed that the environmental risks considered most serious by the general public are different from those considered by the technical professionals charged with reducing environmental risks. Although desirable, it is difficult or almost impossible to change public perception instantly.

Communicating risk is closely associated with public trust and the credibility of the organization, particularly the credibility of the communicator. The public will examine the dedication and competence of the risk communicator closely. They will look at the communicator and wonder how much this person really cares.

The overall scope of risk communication is very broad. This chapter does not attempt to discuss this topic in great detail, but suffice it to say that risk communication is a delicate subject, and communicating risk must be carefully planned. There are a number of areas that can be targeted for effective risk communication. Burke (1992)[227] suggests the following efforts to enhance a risk communication program:

1. developing educational approaches to improve public understanding of the concepts and uncertainties of risk,
2. improving methods to characterize health and environmental risks to policymakers,
3. improving understanding of the risk-information needs of the community, and
4. developing methods to incorporate community values into the risk communication and decision-making process.

Although risk communication by itself will not eliminate conflicts, it is a significant element of a hazardous waste management program. By combining risk communication with a genuine understanding of community values and interests, an important component of improving the decision-making process to address hazardous waste management issues will be provided.

ASSISTANCE IN DEVELOPING PUBLIC PARTICIPATION PLANNING

Developing an effective public participation planning process involves assessing who within the regulatory agency or elsewhere could provide support on public involvement activities and where to obtain information. Most EPA regional offices have one person assigned as the Public Involvement Coordinator (PIC), who serves as a liaison between community members and permit writers, enforcement personnel, facility owners and operators, and other appropriate individuals or groups involved in implementing public participation programs (U.S. EPA, 1993).[228]

The PIC should be able to provide assistance to specific manuals, guidance documents, and memoranda that elaborate on regulations and principles of public participation and to give helpful tips on implementing successful programs. The EPA also provides training in a variety of areas, including public participation, risk communication, and community outreach. If training specific to the RCRA program is not available, usually a planner can easily adapt the community outreach activities used in other programs to the RCRA situation.

Part Five — Hazardous Waste Planning Trends

The last part of this book contains two chapters. Chapter 11 is about the potential to develop a partnership between local and state governments. Because of their proximity to the regulated communities, local governments are in a good position to assist the state in monitoring and enforcing hazardous waste requirements. This chapter examines the procedures, mechanisms, and issues in establishing a state-local government partnership.

There are a number of problems and issues that continue to confront hazardous waste planners in the 1990s. These challenges are discussed in Chapter 12. In this chapter, five major planning issues are briefly described.

11 Hazardous Waste Management by Local Government

A reading of the major environmental protection legislation passed by Congress during the 1970s and 1980s gives us an idea of the respective roles envisioned for the EPA and state and local governments in protecting the environment. The roles were intended to be complementary and supportive, and the allocation of distinct and appropriate responsibilities among sovereigns was the foundation upon which environmental programs were to be built (Hartzog, 1993).[229]

Under this scheme, the EPA has the responsibility for national leadership, oversight, support, and some implementation. National leadership is exercised in the establishment of standards, policies, strategies, financial support, and oversight of program implementation. Delegation of certain environmental protection responsibilities to state and local agencies was envisioned to fully implement the program across the country.

Local government, which includes counties and municipalities, is unique in that it is both a customer and a potential partner of the EPA and the state (U.S. EPA, 1993).[230] As customers, they are considered as regulated parties subject to many of the same environmental statutes, including liability for the cleanup of contamination they may cause. As partners, local governments are often the first line of response to environmental problems caused by the private sector in their jurisdictions.

This chapter is about the potential to develop a partnership between the local governments and the state. Specifically, it examines the partnership in the area of hazardous waste management. The chapter describes existing and possible mechanisms to enhance local involvement, analyzes partnership requirements, and discusses issues in developing a partnership. The next section explains the rationale for developing a state-local partnership.

THE NEED FOR STATE-LOCAL PARTNERSHIPS

Based on the Congressional mandate given to the EPA, each state is responsible for ensuring proper management of hazardous waste generated within its jurisdiction. The assurance is provided through the stringent environmental laws and regulations developed to ensure responsibility of hazardous waste management from cradle-to-grave. The philosophy is manifested in the establishment of strict requirements, such as hazardous waste handling notification, facility permitting, reporting, and record-keeping. The enforcement of these requirements is implemented through inspection and, if necessary, bringing the violators into compliance through litigation.

Inspecting a hazardous waste handler is a time-consuming process. The process consists of reviewing documents related to the handler's background, visually observing the site, documenting the findings, and documenting in writing the results of the inspection. Due to the combination of the large number of handlers in a state and the limited number of state inspectors, most hazardous waste inspections cannot be conducted frequently on the same facility. It is not unusual to discover that a hazardous waste generator received re-inspection from the state environmental agency after five or more years from the previous inspection.

In addition, state inspectors are continuously receiving citizen complaints about improper management of hazardous waste by irresponsible handlers, including unauthorized operation and illegal hazardous waste dumping. Depending on the frequency of these calls, citizen complaints could comprise a large portion of an inspector's time, because investigating one complaint can be as time-consuming as conducting a scheduled inspection.

Confronted with this problem, state regulatory agencies are seeking more effective ways to monitor compliance and enforce applicable regulations. One possible solution is through a partnership with local governments. As mentioned before, because of their proximity to the regulated communities, local governments are in a good position to assist the state in monitoring and enforcing hazardous waste requirements.

Partnering, however, must be seen as an opportunity for local government to address local hazardous waste management problems locally; it must not be interpreted as a mandate from the state or the EPA. To be effective, partnering must look at the benefits and costs of implementing the program at the state level as compared to the local level. It is necessary to decide whether the state's partner is a city, a county, or an association of local governments. This partnering decision must be based on the regionalization scheme of the state.

ESTABLISHING A PARTNERSHIP

The first step in establishing a state-local partnership is to ensure that such a partnership is allowed by state statutes and local ordinances. Once it is determined that the partnership does not violate any laws, the next step is to select a mechanism that can be utilized to facilitate partnering between the state and local governments.

There are a variety of mechanisms that are commonly used by local governments. Local governments may obtain contracts from the state or the EPA, for example, for carrying out inspections, including preparing inspection reports; or they may receive grants for use as primary enforcers of hazardous waste laws and regulations. Local governments may also enter into cooperative agreements for referring information on suspected hazardous waste violations to the state environmental agency, or sign a memorandum of understanding (MOU) with the state environmental agency accepting delegation of authority to perform certain duties and functions specified in the MOU.

Potential duties and functions for local governments in hazardous waste enforcement are the following (U.S. EPA, 1993):[231]

1. Promote compliance through additional outreach to the regulated community.
2. Provide referrals on suspected hazardous waste violations to the state environmental agency.
3. Perform inspections, including the written inspection report.
4. Secure facility compliance with hazardous waste laws and regulations.

The promotion of compliance through outreach is intended to assist facilities' understanding of applicable environmental statutes and regulations. It is not unusual to find a small facility that is not in compliance simply because the owner does not know that the facility is violating environmental laws.

When delegating authority, the state must specify in the MOU what duties and functions are delegated to the local government and what responsibilities are still retained by the state. Examples of delegated functions and duties are hazardous waste generator compliance inspections, citizen complaint investigations, and initial assessment of remedial action needs. Functions and duties that may be retained by the state are, for example, those relating to treatment, storage, and disposal (TSD) facilities, hazardous waste transporters, and the site correction and remediation program.

The MOU must specify personnel qualifications and standard performance levels required for the delegated duties and functions, such as the frequency of inspection for small- or large-quantity generators (SQGs/LQGs). The document also must include reporting procedures of the local government's activities.

Please note that the state-local partnership is not supposed to be limited to the hazardous waste area only. The partnership should be comprehensive and should include all environmental media (air, waste, and water) in which hazardous waste management represents a portion of the overall partnering scheme.

Developing a state-local government partnership is indeed a good and rational undertaking. It will certainly increase the field presence of the regulatory agency, encourage more compliance with environmental laws by the regulated community, and strengthen the program to protect the environment. However, there are a number of issues that may hinder the development of such a partnership. These are discussed in the following section.

ISSUES IN IMPLEMENTING PARTNERSHIPS

Environmental issues are a high priority for local leaders. According to the fifth annual opinion survey of municipal elected officials conducted by the National

League of Cities, among the series of problems facing local governments, the environment ranked first (Fletcher, 1990).[232] When asked to name the two most important problems facing their communities, 23 percent of the local officials surveyed named solid waste. Twenty-one percent cited other environmental problems, giving environmental issues a combined top ranking of 44 percent.

At the same time, local governments in the 1990s will be expected to bear virtually all of the public costs associated with multimedia (air, waste, and water) environmental programs (U.S. EPA, 1990).[233] The EPA estimates that by the year 2000, local governments will bear 87 percent of all public costs for environmental protection — up from 76 percent in 1981. State expenditures will drop from 6 to 5 percent, while EPA spending will decrease from 18 to 8 percent during the 1981–2000 period.

The statistics clearly indicate that the most obvious challenge for local governments in the future is resource constraints. Steinzor (1990)[234] suggests that in order to address this challenge, local governments should develop some loose consensus on which environmental programs should be carried out at the federal level and which at the state and local levels. With resource constraints, carrying out a hazardous waste program at the local level may not be feasible unless the locality is able to obtain resources and expertise to run the program. For example, local governments may run a hazardous waste management program financed by a hazardous waste fee system approved by the local boards or elected officials.

Another issue relates to the designated agency at the local level. This will not be an issue if the municipality or the county has already established or mandated the creation of a local environmental agency. If a local environmental agency is not created, the related duties and functions may be assigned to the existing local agencies. The selection of the local agencies to implement the local environmental program will take time and energy, particularly if, for political reasons, more than one agency are interested in carrying the environmental protection responsibility.

The last issue is about the dichotomy of protecting the environment versus economic development. This dichotomy has always been an issue in the field of hazardous waste management. Similar to the authorization process from the EPA to the state environmental protection agency, the municipality or the county that is authorized to run a portion of the RCRA program in its locality must have a program that is no less stringent than the state RCRA program. Because of this requirement, some local governments may develop RCRA program that is more stringent than that of other municipalities or counties in the state. The more stringent requirements sometimes result in weakening the competitiveness of the respective local government in attracting businesses as compared to its neighboring counties or municipalities. This can ignite friction between the local environmental agency and the local chamber of commerce or economic development agency.

Realizing the potential significant local government problems in partnering with the state and the pressing need to have more field presence in the implementation of environmental mandates, some states have expanded their program geographically by establishing satellite state-operated offices throughout the state.

The creation of the regional offices is intended to better serve the customers throughout the state. The regulatory function of the regional offices is normally field-oriented; i.e., they routinely inspect regulated facilities, handle citizen complaints, and respond to emergency spill calls in their jurisdictions. It is important to underline that regional offices can be used with or without partnering.

12 The Future of Hazardous Waste Planning

Managing hazardous waste in the 1990s represents a unique challenge for hazardous waste planners and managers. This is indicated by the variety of problems and issues described in Chapter 1. These issues are chronic and will continue to exist as long as people generate hazardous waste.

The hazardous waste regulations, for example, will continue to be stricter. The EPA will continue to emphasize enforcement and impose higher penalties for violations of hazardous waste laws. The number of fines, civil penalties, and criminal cases will rise under the EPA's reorganized enforcement structure (Air/Water Pollution Report, 1994).[235]

The integration of a cross-program and multimedia approach (i.e., involving components of air, waste, and water) into all stages of the inspection and enforcement program will be accepted as the norm. This multimedia approach will enable the regulatory agency to better pursue its goal of protecting public health and the environment through deterrence while maintaining an efficiency that could not be achieved through the use of traditional single-media approaches alone (U.S. EPA, 1993).[236]

The "Not in my backyard" (NIMBY) syndrome will not go away, and anger over what community activists perceive as a pattern of locating industries near poor, minority residences has awakened the EPA, Congress, and the industrial community to a new societal issue: *environmental justice.* Initially a local concern, environmental justice has now garnered broader national support (Environment Reporter, 1994).[237]

Environmental justice, also known as environmental equity, can be divided into three major categories (Bullard, 1991):[238]

- **Procedural equity**, which questions whether governments apply rules, regulations, and evaluation criteria uniformly across the board. Inequity occurs, for example, by excluding persons from decision-making boards, commissions, and committees; by holding hearings in remote locations and at inconvenient times in order to minimize public participation; and by writing notices and conducting public meetings using English in predominantly non-English-speaking communities.

- **Geographic equity**, which occurs when a neighborhood, community, or region is disproportionately burdened by environmental threats such as hazardous waste.
- **Social equity**, which deals with the role of race and class in environmental decision-making.

An awareness of environmental justice will ultimately lead to the reduction or prevention of harmful exposure to hazardous wastes at the community level (Harding and Holdren Jr., 1993).[239] Harding and Holdren predict that the mechanism for reducing exposures from future industrial practices will come primarily via increased waste minimization and pollution prevention efforts. Environmental justice will also become one of many criteria that scientists and the public will be asked to use in assessing, preventing, and remediating environmental hazards. In this case, Harding and Holdren argue that although it is true that the assessment of risk requires environmental professionals to maintain a strong technical foundation, technical capability per se will no longer be sufficient to address the diversity of needs and concerns that communities have regarding responsible waste management practices.

The environmental justice issue will force the hazardous waste industry to make some adjustments in such things as facility siting and cleanup standards (Hazardous Waste News, 1994).[240] While the cleanup standard for remediating a site will trigger a forum of continuing debate, the number of newly found sites contaminated with hazardous waste will become larger than the number of sites remedied. This situation will stimulate the development of more effective and efficient site correction and remediation planning.

The hazardous waste management costs will continue to be expensive. The data quality issue, such as the quantification problem in measuring the success of waste minimization and pollution prevention, will still not be easy to resolve. The data quality issue is also applicable to the waste generation data.

The issue of balance between protecting the environment and promoting economic development is prevalent; it is constantly examined and debated. Traditionally, it is believed that the areas with the weakest environmental regulations are expected to have the strongest economies. On this subject, a Massachusetts Institute of Technology (MIT) economist testified at a Congressional staff briefing on September 21, 1993, that states with the strongest environmental regulations appear to have the strongest overall economies (Environment Reporter, 1993).[241] When plotting regulations versus growth factors for each state, Professor Stephen M. Meyer found a positive or no relationship between regulation and economic growth.

Professor Meyer's study, which was based on a review of 20 years of economic data, concluded that the level of a state's environmental regulation does not have a lasting or observable impact on economic growth. He observed, however, that recessions may penalize states that have tough environmental laws and may lead to lower employment. These states, on the other hand, were the first to break from the recessions as the economy improved.

The preceding discussions indicate that the top five issues that will dominate the planning and management of hazardous waste in the 1990s are:

1. the integration of environmental justice into program planning and management,
2. the continuing promotion of waste minimization and pollution prevention,
3. the escalation of multimedia enforcement,
4. the acceleration of the cleanup of contaminated sites, and
5. the determination of the cleanup standard, including the discussion of risk assessment and comparative risk.

Additionally, there is a continuing need for making hazardous waste regulations more understandable. Better understanding of the regulations will increase compliance with the laws, reduce the costs to comply, and help make enforcement more efficient and effective. Better knowledge of risk and science must be behind hazardous waste management standards, and governmental hazardous waste programs must be built on a strong advisory role to help industries help themselves. These are just a few examples of concerns that still characterize the 1990s.

Parallel to the growing problems and issues in the hazardous waste management field, the role of hazardous waste planners will also continue to grow. It can be anticipated that the profession will play a more significant role than it has in the past as better planning and management of hazardous waste is continuously demanded by the public.

Notes

1. U.S. Environmental Protection Agency, *Enforcement Accomplishments Report,* U.S. EPA, Washington, D.C., 1990.
2. U.S. General Accounting Office, *Penalties May Not Recover Economic Benefits,* U.S. GAO, Washington, D.C., 1991.
3. "Court Fines Bethlehem Steel $6 Million, Imposes Record Penalty for RCRA Violations," *Environment Reporter,* September 10, 1993, pp. 845–846.
4. "Mississippi Man Sentenced to Prison for Illegal Dumping of Hazardous Waste," *Environment Reporter,* March 30, 1990, p. 1918.
5. "Civilians Ask Supreme Court to Review RCRA Convictions at Army Weapons Facility," *Environment Reporter,* December 14, 1990, pp. 1556–1557.
6. "Three Federal Managers Convicted of HazWaste Disposal Violations," *Waste Tech News,* March 13, 1989, pp. 1 and 14.
7. American for the Environment, *The Rising Tide: Public Opinion, Policy and Politics,* American for the Environment, Washington, D.C., 1989.
8. Popper, Frank, "The Environmentalist and the LULU," *Environment,* Volume 27, Number 2, 1985, pp. 7–11 and 37–40.
9. Heiman, Michael, "From 'Not in My Backyard!' to 'Not in Anybody's Backyard!'" *Journal of the American Planning Association,* Volume 56, Number 3, 1990, pp. 359–362.
10. New York State Legislature, *Hazardous Waste Facility Siting: A National Survey,* Commission on Toxic Substances and Hazardous Wastes, New York, NY, 1987.
11. Brown, Michael, *Laying Waste: The Poisoning of America by Toxic Chemicals,* Pantheon Books, New York, NY, 1980.
12. Silverman, Gerald, "Love Canal: A Retrospective," *Environment Reporter,* September 15, 1989, pp. 835–850.
13. "Reilly Endorses Love Canal Study Finding Two-Thirds of Area Habitable," *Environment Reporter,* May 18, 1990, pp. 191–192.
14. Silverman, Gerald, "Resettlement of Love Canal Begins, But Bank Sour on Mortgage," *Environment Reporter,* December 21, 1990, pp. 1590–1592.
15. Vig, Norman, and Kraft, Michael, *Environmental Policy in the 1980s: Reagan's New Agenda*, Congressional Quarterly Inc., Washington, D.C., 1983.
16. "Times Beach, Cleanup Begins in Abandoned Town," *Superfund Report,* May 8, 1991, p. 20.
17. DiPasquale, Suzanne, "Times Beach, Judge Approves Plan to Incinerate Dioxin-Tainted Soil," *Superfund Report,* January 16, 1991, p. 22.

18. "Settlement Reached in Cleanup of Times Beach: Syntex Agribusiness Agrees to Pay $10 million," *Environment Reporter,* July 27, 1990, pp. 531–532.
19. U.S. Environmental Protection Agency, *Environmental Progress and Challenges: EPA's Updates,* U.S. EPA, Washington, D.C., 1988.
20. "533 PRPs Sign $22.4M Pact for Michigan Site Cleanup," *Waste Tech News,* June 5, 1989, p. 5.
21. U.S. Environmental Protection Agency, *Enforcement Accomplishments Report,* U.S. EPA, Washington, D.C., 1990.
22. "EPA Reveals Largest Superfund Consent Decree," *Waste Tech News,* September 10, 1990, pp. 1 and 15.
23. U.S. Environmental Protection Agency, *Enforcement Accomplishments Report,* U.S. EPA, Washington, D.C., 1990.
24. Forcella, Domenic, *The Role of Waste Minimization,* National Governors' Association, Washington, D.C., 1989.
25. U.S. Environmental Protection Agency, *Land Disposal Restrictions, Summary of Requirements,* U.S. EPA, Washington, D.C., 1989.
26. U.S. General Accounting Office, *Nonhazardous Waste: Environmental Safeguard for Industrial Facilities Need to be Developed,* U.S. GAO, Washington, D.C., 1990.
27. "New Toxicity Rule to Cost $250M–$400M Annually," *Waste Tech News,* March 26, 1990, pp. 1 and 13.
28. Morgulec, James A., and Woitte, Deborah K., "Resource Conservation and Recovery Act," *Environmental Crime Manual,* U.S. Department of Justice, Washington, D.C., 1991, pp. VII.1–VII.78.
29. Environmental Information Ltd., *Interdependence in the Management of Hazardous Waste,* EI Ltd., Minneapolis, MN, 1993.
30. Cobb, Susan, "Justice Department Opposes Alabama Waste Ban," *Waste Tech News,* March 12, 1990, pp. 1 and 16.
31. "Governor Signs Hazardous Waste Import Bill: Owners of Commercial Landfill Plan Lawsuit," *Environment Reporter,* June 22, 1990, pp. 376–377.
32. "Interstate Waste Transport — Appeals Court Finds Barrier Unconstitutional," *Superfund Report,* August 15, 1990, pp. 2 and 16.
33. Environmental Information Ltd., *Interdependence in the Management of Hazardous Waste,* EI Ltd., Minneapolis, MN, 1993.
34. Hilton, Cynthia, "States Slap Fees on HazWaste Transport," *Waste Age,* May 1989, pp. 69–71.
35. "Decision Rejecting Alabama Waste Law Will Not Be Reviewed, Supreme Court Says," *Environment Reporter,* June 14, 1991, p. 338.
36. "Alabama's Holley Bill Limiting Shipment of Hazardous Waste Into the State Ruled Unconstitutional," *National Environmental Enforcement Journal,* October 1990, pp. 18–20.
37. "Interstate Waste Transport — Appeals Court Finds Barrier Unconstitutional," *Superfund Report,* August 15, 1990, pp. 2 and 16.
38. Becker, W. Virginia, *Legal Issues Affecting Interstate Disposal,* National Governors' Association, Washington, D.C., 1989.
39. U.S. General Accounting Office, *Hazardous Waste: Uncertainties of Existing Data,* U.S. GAO, Washington, D.C., 1987.
40. U.S. General Accounting Office, *SARA Capacity Assurance: Data Problems Underlying the 1989 State Assessments,* U.S. GAO, Washington, D.C., 1991.

41. U.S. General Accounting Office, *EPA's Generation and Management Data Need Further Improvement,* U.S. GAO, Washington, D.C., 1990.
42. The delegation of authority from the EPA to states is discussed in Chapter 4.
43. U.S. Environmental Protection Agency, *RCRA Oriental Manual 1990 Edition,* U.S. EPA, Washington, D.C., 1990.
44. The role of local government in hazardous waste management is discussed in Chapter 11.
45. Lindgren, Gary, *Managing Industrial Hazardous Waste, A Practical Handbook,* Lewis Publishers, Chelsea, MI, 1989.
46. "EPA Grants Three-Month Stay of Standards for Third-Third Wastes in Final Land Ban Rule," *Environment Reporter,* November 5, 1990, pp. 147–150.
47. Lowrance, Sylvia, "EPA Completes Land Disposal Restrictions Program for Hazardous Wastes with Third Third Rule," *Waste Age,* September 1990, pp. 38–41 and 176.
48. Seymour, John, "Civil and Criminal Liability of Corporate Officers under Federal Environmental Laws," *Environment Reporter,* June 9, 1989, pp. 337–348.
49. McGregor, Gregor, "Managing for Environmental Protection," *Waste Tech News,* January 16, 1989, pp. 10 and 12.
50. "Environmental Decade," *World Waste,* January 1990, p. 6.
51. McDowell, Bruce D., "Approaches to Planning," in Frank S. So et al., Eds., *The Practice of State and Regional Planning,* American Planning Association, Washington, D.C., 1986, pp. 3–22.
52. DeBettencourt, J. S. et al., "Making Planning More Responsive to its Users: The Concept of Metaplanning," *Environment and Planning,* Volume A, Number 14, 1982, pp. 311–322.
53. LeBreton, Preston, and Henning, Dale, *Planning Theory,* Prentice Hall, Englewood Cliffs, NJ, 1961.
54. Buskirk, Drannon, "Environmental Impact Analysis," in Frank S. So et al., Eds., *The Practice of State and Regional Planning,* American Planning Association, Washington, D.C., 1986, pp. 238–254.
55. Thibodeaux, Louis, "Hazardous Material Management in the Future," *Environmental Science and Technology,* Volume 24, Number 4, 1990, pp. 456–459.
56. Rowe, P. et al., *Principles for Local Environmental Management,* Ballinger, Cambridge, MA, 1976.
57. Baldwin, John, *Environmental Planning and Management,* Westview Press, Boulder, CO, 1985.
58. Daley, Peter, "Comprehensive Hazardous Waste Treatment Facilities," in Harry Freeman, Ed., *Standard Handbook of Hazardous Waste Treatment and Disposal,* McGraw-Hill, New York, NY, 1989, pp. 11.1–11.18.
59. Korte, F., "Introductory General Statement on Hazardous Wastes in the Environment," in Richard Abbou, Ed., *Hazardous Waste Detection, Control, Treatment,* Elsevier, New York, NY, 1988, pp. xix–xx.
60. Maltezou, Sonia, Biswas, Asit, and Sutter, Hans, Eds., *Hazardous Waste Management,* Tycooly, New York, NY, 1989.
61. Tedder, William, and Pohland, Frederick, Eds., *Emerging Technologies in Hazardous Waste Management,* American Chemical Society, Washington, D.C., 1990.
62. Campbell, Monica, and Glenn, William, *Profit from Pollution Prevention,* Pollution Probe Foundation, Toronto, Canada, 1982.

63. Metry, Amir, "Elements of a Balanced Approach Towards Hazardous Waste Management," in Sonia Maltezou et al., Eds., *Hazardous Waste Management,* Tycooly, New York, NY, 1989.
64. Lindgren, Gary, *Managing Industrial Hazardous Waste, A Practical Handbook,* Lewis Publishers, Chelsea, MI, 1989.
65. Hunt, Gary, and Schecter, Roger, "Minimization of HazWaste Generation," in Harry Freeman, Ed., *Standard Handbook of HazWaste Treatment and Disposal,* McGraw-Hill, New York, NY, 1989, pp. 5.1–5.26.
66. Martin, Larry, *Proven Profits from Pollution Prevention,* Institute for Local Self-Reliance, Washington, D.C., 1989.
67. Zosel, Thomas, "Case Study: How 3M Makes Pollution Prevention Pay Big Dividends," *Pollution Prevention Review,* Volume 1, Number 1, 1990, pp. 67–72.
68. Hansen, Fred, "Pollution Prevention, A New Mandate for Oregon's Environment," *The Environmental Forum,* September/October 1989, pp. 30–34.
69. Shorthouse, Barry, "Using Risk Analysis to Set Priorities for Pollution Prevention," *Pollution Prevention Review,* Volume 1, Number 1, 1990, pp. 41–53.
70. U.S. Environmental Protection Agency, *Waste Minimization Opportunity Assessment,* U.S. EPA, Washington, D.C., 1988.
71. Wolf, Katy, "Source Reduction and the Waste Management Hierarchy," in Harry Freeman, Ed., *Hazardous Waste Minimization,* Air and Waste Management Association, Pittsburgh, PA, 1989, pp. 107–112.
72. Theodore, Louis, and McGuinn, Young C., *Pollution Prevention,* Van Nostrand Reinhold, New York, NY, 1992.
73. U.S. General Accounting Office, *Hazardous Waste: New Approach Needed to Manage the Resource Conservation and Recovery Act,* U.S. GAO, Washington, D.C., 1988.
74. U.S. Environmental Protection Agency, *Report to Congress: Minimization of Hazardous Waste,* U.S. EPA, Washington, D.C., 1986.
75. U.S. Environmental Protection Agency, *Environmental Progress and Challenges: EPA's Updates,* U.S. EPA, Washington, D.C., 1988.
76. Tanaka, Masaru, "Strategy for Hazardous Waste Management," in K. Zirm and J. Mayer, Eds., *The Management of Hazardous Substances in the Environment,* Elsevier, New York, NY, 1990.
77. Maltezou, Sonia, Biswas, Asit, and Sutter, Hans, Eds., *Hazardous Waste Management,* Tycooly, New York, NY, 1989.
78. Wentz, Charles, *Hazardous Waste Management,* McGraw-Hill, New York, NY, 1989.
79. Sutter, Hans, "Review of Hazardous Waste Management Systems as Applied by the Government and Private Sectors," in Sonia Maltezou et al., Eds., *Hazardous Waste Management,* Tycooly, New York, NY, 1989, pp. 3–23.
80. Lindgren, Gary, *Managing Industrial Hazardous Waste, A Practical Handbook,* Lewis Publishers, Chelsea, MI, 1989.
81. Fox, Edward Z., "Educating Decision Makers in Pollution Control," *Pollution Prevention Review,* Volume 1, Number 1, 1990, pp. 3–12.
82. Sell, Nancy J., *Industrial Pollution Control: Issues and Techniques,* Van Nostrand Reinhold, New York, NY, 1992.
83. Toner, William, "Environmental Land Use Planning," in Frank S. So and Judith Getzels, Eds., *The Practice of Local Government Planning,* International City Management Association, Washington, D.C., 1988, pp. 117–138.

84. Slater, David, *Management of Local Planning,* International City Management Association, Washington, D.C., 1984.
85. Case, David R., "Resource Conservation and Recovery Act," in *Environmental Law Handbook,* Government Institutes Inc., Rockville, MD, 1993, pp. 60–93.
86. Smith, Herbert H., *The Citizen's Guide to Zoning,* American Planning Association, Chicago, IL, 1983.
87. U.S. Environmental Protection Agency, *RCRA Orientation Manual 1990 Edition,* U.S. EPA, Washington, D.C., 1990.
88. Some states may require annual reports instead of biennial reports. By imposing this requirement, state rules are more stringent than federal regulations. Under RCRA, more stringent state rules supersede federal regulations.
89. Wentz, Charles A., *Hazardous Waste Management,* McGraw-Hill, New York, NY, 1992.
90. "Corrective Action — How It Works," *Superfund Report,* July 18, 1990, p. 3.
91. Davis, Robert C., "RCRA Corrective Action Developments," in Kenneth M. Kastner, Ed., *Current Developments in RCRA,* Government Institutes Inc., Rockville, MD, 1993, pp. 184–201.
92. Hope, Michael R., "RCRA/CERCLA Interaction," in Kenneth M. Kastner, Ed., *Current Developments in RCRA,* Government Institutes Inc., Rockville, MD, 1993, pp. 213–223.
93. Hall Jr., Ridgway M. et al., *RCRA Hazardous Wastes Handbook,* Ninth Edition, Government Institutes Inc., Rockville, MD, 1991.
94. Lee, Robert T., "Comprehensive Environmental Response, Compensation, and Liability Act," in *Environmental Law Handbook,* Government Institutes Inc., Rockville, MD, 1993, pp. 267–320.
95. Reinhardt, John R., "Summary of Resource Conservation and Recovery Act Legislation and Regulation," in Harry M. Freeman, Ed., *Standard Handbook of Hazardous Waste Treatment and Disposal,* McGraw-Hill, New York, NY, 1989, pp. 1.9–1.26.
96. Quarles, John, *Federal Regulations of Hazardous Wastes, A Guide to RCRA,* Environmental Law Institute, Washington, D.C., 1982.
97. Green, Carol L., Dunn, James R., and Oppenheimer, Peter H., "RCRA Enforcement in the '90s: Avoiding Civil and Criminal Liability," in Kenneth M. Kastner, Ed., *Current Developments in RCRA*, Government Institutes Inc., Rockville, MD, 1993, pp. 75–110.
98. Briggum, Sue M. et al., *Hazardous Waste Regulation Handbook, A Practical Guide to RCRA and Superfund,* Revised Edition, Executive Enterprises Publications Co. Inc., New York, NY, 1985.
99. U.S. Environmental Protection Agency, *RCRA Authorization and Joint Permitting in Authorized States,* U.S. EPA, Washington, D.C., 1985.
100. Hall Jr., Ridgway M. et al., *RCRA Hazardous Wastes Handbook,* Ninth Edition, Government Institutes Inc., Rockville, MD, 1991.
101. Gruber, William, "State Program Status, An Update of RCRA Authorization," *EI Digest,* January 1991, pp. 34–48.
102. U.S. Environmental Protection Agency, *State Consolidated RCRA Authorization Manual,* U.S. EPA, Washington, D.C., 1988.
103. U.S. Environmental Protection Agency, *State Authorization Manual,* U.S. EPA, Washington, D.C., 1990.
104. McCoy and Associates Inc., *RCRA Regulations and Keyword Index,* McCoy and Associates Inc., Lakewood, CO, 1992.

105. Morgulec, James A., and Woitte, Deborah K., "Resource Conservation and Recovery Act," in *Environmental Crimes Manual,* U.S. Department of Justice, Washington, D.C., 1991, pp. VII.1–VII.78.
106. J. J. Keller and Associates Inc., *Hazardous Waste Regulatory Guide,* J. J. Keller and Associates Inc., Neenah, WI, 1993.
107. Patton, Carl V., "Information for Planning," in Frank S. So and Judith Getzels, Eds., *The Practice of Local Government Planning,* International City Management Assn., Washington, D.C., 1988, pp. 472–502.
108. Wagner, Travis P., *Hazardous Waste Identification and Classification Manual,* Van Nostrand Reinhold, New York, NY, 1990.
109. U.S. Environmental Protection Agency, *1993 Hazardous Waste Report — Instructions and Forms,* U.S. EPA, Washington, D.C., 1993.
110. Morgulec, James A., and Woitte, Deborah K., "Resource Conservation and Recovery Act," in *Environmental Crimes Manual,* U.S. Department of Justice, Washington, D.C., 1991, pp. VII.1–VII.78.
111. U.S. General Accounting Office, *Hazardous Waste, EPA's Generation and Management Data Need Further Improvement,* U.S. GAO, Washington, D.C., 1990.
112. Cheremisinoff, Paul N., "Selecting Hazardous Waste Treatment Systems: Thermal Classification is Key," in *The National Environmental Journal,* Volume 4, Issue 1, 1994, pp. 49–51.
113. Marvin, Richard, "Hazardous Waste Tracking Issues," in *HazMat World,* Volume 6, Number 8, 1993, pp. 56–57.
114. National Governors' Association, *A State Guide to the Uniform Hazardous Waste Manifest,* NGA, Washington, D.C., 1991.
115. U.S. Environmental Protection Agency, *Guidance for Capacity Assurance Planning — Capacity Planning Pursuant to CERCLA §104(c)(9),* U.S. EPA, Washington, D.C., 1993.
116. Soesilo, Joseph A., and Wilson, Stephanie R., *Arizona Industrial Waste — 1987 Profile,* Arizona Department of Environmental Quality, Phoenix, AZ, 1989.
117. The secondary waste generation can be analyzed through the review of origin code provided in the GM Form.
118. Soesilo, Joseph A., and Wilson, Stephanie R., *Arizona Waste Minimization Practice,* Arizona Department of Environmental Quality, Phoenix, AZ, 1990.
119. Soesilo, Joseph A., "Measuring State Hazardous Waste Flow and the Policy Impacts of Different Hazardous Waste Management Rules Between Neighboring States," in *Proceedings of Papers, International Association for Impact Assessment,* I.A.I.A., Lausanne, Switzerland, 1990, pp. 266–268.
120. U.S. Environmental Protection Agency, *Biennial Reporting System (BRS) Training,* U.S. EPA, Washington, D.C., 1992.
121. Focus Information Builders Inc., *PC/FOCUS — Getting Started,* Focus Information Builders Inc., New York, NY, 1989.
122. U.S. Environmental Protection Agency, *Biennial Reporting System (BRS) User Guide Version 2.1.0,* U.S. EPA, Washington, D.C., 1992.
123. U.S. Environmental Protection Agency, *Fundamentals of Environmental Compliance Inspections — Third Printing,* Government Institutes Inc., Rockville, MD, 1993.
124. Hopkins, Michael, "Compliance — A Hot Issue in Industry," in *Waste Tech News,* Volume 5, Number 16, 1993, pp. 1 and 3.

125. Lindgren, Gary F., *Managing Industrial Hazardous Waste — A Practical Handbook,* Lewis Publishers, Chelsea, MI, 1989.
126. U.S. Environmental Protection Agency, *Resource Conservation and Recovery Act Inspection Manual — Second Edition,* Government Institutes Inc., Rockville, MD, 1992.
127. Cahill, Lawrence B., and Kane, Raymond W., Eds., "Typical Compliance Problems Found During Audits," in *Environmental Audits, Sixth Edition,* Government Institutes Inc., Rockville, MD, 1992, pp. X-1 through X-43.
128. U.S. Environmental Protection Agency, "Environmental Auditing Policy Statement," *51 Federal Register 25004,* July 9, 1986.
129. Green, Carol L., Dunn, James R., and Oppenheimer, Peter H., "RCRA Enforcement in the '90s: Avoiding Civil and Criminal Liability," in Kenneth M. Kastner, Ed., *Current Developments in RCRA,* Government Institutes Inc., Rockville, MD, 1993, pp. 75–110.
130. Kuusinen, Tapio, "A Government Perspective," in Lawrence B. Cahill and Raymond W. Kane, Eds., *Environmental Audits — Sixth Edition,* Government Institutes Inc., Rockville, MD, 1992, pp. II-1 through II-14.
131. Shanoff, Barry, "Using Self-Audit Data to Improve Compliance," in *World Wastes,* June 1993, p. 22.
132. Levin, Michael H., Hymes, Allan D., and Mullaney, Sean, "Discovery and Disclosure: How to Protect Your Environmental Audit Report," in *Environment Reporter,* Volume 24, Number 36, 1994, pp. 1606–1612.
133. Hirschborn, Joel S., "Economics of Waste Reduction, Resource Recovery, and Recycling," in Thomas F. P. Sullivan, Ed., *The Greening of American Business,* Government Institutes Inc., Rockville, MD, 1992, pp. 144–159.
134. U.S. Environmental Protection Agency, *Report to Congress,* U.S. EPA, Washington, D.C., 1986.
135. U.S. General Accounting Office, *Toxic Substance: Advantages of and Barriers to Reducing the Use of Toxic Chemicals,* U.S. GAO, Washington, D.C., 1992.
136. Anderson, Steven, and Herb, Jeanne, "Building Pollution Prevention into Facilitywide Permitting," in *Pollution Prevention Review,* Volume 2, Number 4, 1992, pp. 415–430.
137. Foecke, Terry, and Style, Robert, "An Update on State Facility Planning Legislation and Programs," in *Pollution Prevention Review,* Volume 2, Number 4, 1992, pp. 479–490.
138. Ohio Environmental Protection Agency, *Facility Pollution Prevention Planning,* OEPA, Columbus, OH, 1991.
139. U.S. Environmental Protection Agency, *Waste Analysis at Facilities that Generate, Treat, Store, and Dispose of Hazardous Waste — A Guidance Manual,* U.S. EPA, Washington, D.C., 1992.
140. U.S. Environmental Protection Agency, *Land Disposal Restrictions — Summary of Requirements,* U.S. EPA, Washington, D.C., 1991.
141. The wastes are known as the "California List" wastes.
142. Case, David R., "Resource Conservation and Recovery Act," in *Environmental Law Handbook,* Government Institutes Inc., Rockville, MD, 1993, pp. 60–93.
143. Graham, Ann B., "The Results of PPR's 1993 Survey: Industry's PP Practices," in *Pollution Prevention Review,* Volume 3, Number 4, 1993, pp. 369–381.
144. INFORM, *Environmental Dividends: Cutting More Chemical Wastes,* INFORM, New York, NY, 1992.

145. Minnesota Office of Waste Management, *Minnesota Guide to Pollution Prevention Planning,* MOWM, St. Paul, MN, 1991.
146. U.S. Environmental Protection Agency, *Facility Pollution Prevention Guide,* U.S. EPA, Washington, D.C., 1992.
147. Arizona Department of Environmental Quality, *Pollution Prevention Plan Guidance Manual,* ADEQ, Phoenix, AZ, 1992.
148. Cahill, Lawrence B., and Kane, Raymond W., Eds., *Environmental Audits,* Government Institutes Inc., Rockville, MD, 1992.
149. Higgins, Thomas E., *Hazardous Waste Minimization Handbook,* Lewis Publishers, Chelsea, MI, 1991.
150. Ferrante, Louise M., "Preparing to Meet Waste Minimization Mandates," in *The National Environmental Journal,* November/December 1993, pp. 55–58.
151. Rozell, David K., and Brower, Roy W., "Pollution Prevention Facility Planning: The Oregon Experience," in *Pollution Prevention Review,* Volume 3, Number 3, 1993, pp. 277–283.
152. Baker, Rachel D., Dunford, Richard W., and Warren, John L., *Alternatives for Measuring Hazardous Waste Reduction,* Hazardous Waste Research and Information Center, Champaign, IL, 1991.
153. U.S. Environmental Protection Agency, "Notice of Availability of Pollution Prevention Grants and Announcement of Financial Assistance Programs Eligible for Review," *58 Federal Register 16664,* March 30, 1993.
154. States are encouraged but not mandated to use the BRS as a source of information for their CAPs.
155. U.S. Environmental Protection Agency, *Using Tabletalk to Prepare CAP Tables,* U.S. EPA, Washington, D.C., 1992.
156. U.S. Environmental Protection Agency, *Guidance for Capacity Assurance Planning — Capacity Planning Pursuant to CERCLA §104(c)(9),* U.S. EPA, Washington, D.C., 1993.
157. The capacity of any treatment facility is measured annually (i.e., tons per year), because the capacity is assumed constant over the years. For a landfill, the capacity is not constant but depleting every year, therefore the landfill capacity is measured by its availability of capacity in the observed year.
158. U.S. Environmental Protection Agency, *Guidance for Capacity Assurance Planning — Capacity Planning Pursuant to CERCLA §104(c)(9),* U.S. EPA, Washington, D.C., 1993.
159. U.S. Environmental Protection Agency, *Guidance for Capacity Assurance Planning — Draft One-Time Waste Estimates for Capacity Assurance Planning,* U.S. EPA, Washington, D.C., 1994.
160. U.S. General Accounting Office, *SARA Capacity Assurance: Data Problems Underlying the 1989 State Assessment,* U.S. GAO, Washington, D.C., 1991.
161. Soesilo, J. Andy, *Estimation of Household Hazardous Waste,* paper presented to the Governor's Hazardous Waste Disposal Committee, Arizona Department of Environmental Quality, Phoenix, AZ, 1992.
162. Kusterer, Thomas, "Landfill Siting in a High-Growth Metropolitan County," in *Environmental Decision,* February 1991, pp. 4–5 and 21–22.
163. Apogee Research Inc., *Analysis of Siting Attempts and the Factors That Influence Success,* National Governors' Association, Washington, D.C., 1988.
164. Houghton, Mary J., *Siting New Treatment and Disposal Facilities,* National Governors' Association, Washington, D.C., 1989.

165. Apogee Research Inc., *Analysis of Siting Attempts and the Factors That Influence Success,* National Governors' Association, Washington, D.C., 1988.
166. Becker, Victoria W., *Siting Hazardous Waste Facilities: Guiding Principles,* National Governors' Association, Washington, D.C., 1993.
167. Sanderson, Debra R., "Negotiating Facility Siting Remedies," in *Western City,* August 1985.
168. Bingham, Gail, and Miller, Daniel S., "Prospects for Resolving Hazardous Waste Siting Disputes Through Negotiation," in *Natural Resources Lawyer,* Fall 1984, pp. 1–17.
169. Daly, John B., and Vitaliano, Eric N., *Hazardous Waste Facility Siting in New York State, the Evolution of a Promising Public Policy,* New York State Joint Legislative Committee on Toxic Substance and Hazardous Wastes, New York, NY, 1989.
170. Gregory, Robin, and Kunreuther, Howard, "Successful Siting Incentives," in *Civil Engineering,* April 1990, pp. 73–75.
171. U.S. Environmental Protection Agency, *Guidance for Capacity Assurance Planning — Capacity Planning Pursuant to CERCLA §104(c)(9),* U.S. EPA, Washington, D.C., 1993.
172. Environmental Resources Management, *Report on the National Conference on State Capacity Assurances,* Consortium of State Hazardous Waste Siting Authorities, Boston, MA, 1989.
173. U.S. Environmental Protection Agency, *Hazardous Waste Management Planning Needs and Practices — A Review of Several State Agency Approaches,* U.S. EPA, Washington, D.C., 1993.
174. Washington State Department of Ecology, *Washington State Hazardous Waste Plan,* WSDOE, Olympia, WA, 1992.
175. U.S. Environmental Protection Agency, *The Nation's Hazardous Waste Management Program at a Crossroads,* U.S. EPA, Washington, D.C., 1990.
176. U.S. Environmental Protection Agency, *National Biennial RCRA Hazardous Waste Report — Based on 1989 Data,* U.S. EPA, Washington, D.C., 1993.
177. Von Oppenfeld, Rolf R., "RCRA Implications of Documented Hazardous Waste Treatment, Storage, or Disposal at Facilities Investigated in Due Diligence Transactions or in Environmental Audits: The Obligations to Notify, Permit, and Remediate for Audited Facilities," in *HazWaste 93 Symposium,* Phoenix, AZ, 1993, pp. 1–23.
178. In the RCRA program, a business that generates the regulated quantity of hazardous waste is subject to the generator requirements; the business is not required to obtain an RCRA permit unless it treats, disposes on-site, or stores the hazardous waste for more than 90 days.
179. Lee, Robert T., "Comprehensive Environmental Response, Compensation, and Liability Act," in *Environmental Law Handbook,* Government Institutes Inc., Rockville, MD, 1993, pp. 267–320.
180. Davis Jr., Robert C., "RCRA Corrective Action Developments," in Kenneth M. Kastner and Bryan Cave, Eds., *Current Developments in RCRA,* Government Institutes Inc., Rockville, MD, 1993, pp. 184–201.
181. McCarroll, John, "Introduction," in *Corrective Action Training Course,* U.S. EPA Region–IX, San Francisco, CA, 1993.
182. Corbaley, Susan, "Corrective Action and the RCRA Facility Assessment Program," in *RCRA Facility Assessment Training for Arizona Department of Environmental Quality,* Science Applications International Corporation, San Francisco, CA, 1993.

183. Wesling, Mary, "Conducting the Visual Site Inspection," in *RCRA Facility Assessment Training for Arizona Department of Environmental Quality,* Science Applications International Corporation, San Francisco, CA, 1993.
184. Corbaley, Susan, "Preparing the RFA Report Under the RCRA Facility Assessment Program," in *RCRA Facility Assessment Training for Arizona Department of Environmental Quality,* Science Applications International Corporation, San Francisco, CA, 1993.
185. Saracino, Ray, "RCRA Facility Investigation," in *Corrective Action Training Course 1993,* U.S. Environmental Protection Agency, Region 9, San Francisco, CA, 1993.
186. Soesilo, Joseph A., and Curry, Thomas E., "A Comprehensive Approach to Addressing RCRA Contaminated Sites" (draft), 1994.
187. Upon discovery of contamination, RCRA Subtitle C requires a facility to perform cleanup immediately. In this context, the term *immediate* is not precisely defined by the EPA. This flexibility enables the state agency to approve voluntary cleanup without imposing any permit requirements.
188. Power, Jan G., and Leifer, Steven, "Expediting Site Cleanups," in *Waste Age,* June 1993, pp. 177–184.
189. Under the 1984 HSWA, the EPA must require corrective action for all releases of hazardous waste to facilities seeking closure or post-closure permits as well as those seeking ordinary operating permits.
190. The EPA states that if a facility fails to utilize the opportunity to immediately clean up the contamination, it must obtain an RCRA permit or conduct a cleanup under a "comparable authority," i.e., through consent or compliance orders (55 FR 30807).
191. Travis, Curtis C., and Doty, Carolyn B., "Remedial Action Decision Process," in Richard B. Gammage and Barry A. Berven, Eds., *Hazardous Waste Site Investigations — Toward Better Decisions,* Lewis Publishers, Boca Raton, FL, 1992, pp. 261–270.
192. Ronan III, John T., "New Horizons for Superfund Enforcement," in *HazMat World,* February 1990, pp. 42–54.
193. Carr, F. Housley, "What a Dump!" in *Resources,* January 1991, pp. 12–14.
194. "Lautenberg, Dingell Blast Superfund Studies; Reilly Focuses on Accomplishments of Program," in *Environment Reporter,* October 11, 1991, p. 1531.
195. Burke, Thomas A., "Overview: Refining Hazardous Waste Site Policies Through Research," in Richard B. Gammage and Barry A. Berven, Eds., *Hazardous Waste Site Investigations — Toward Better Decisions,* Lewis Publishers, Boca Raton, FL, 1992, pp. 221–225.
196. Russell, Milton, Colglazier, E. William, and English, Mary R., *Hazardous Waste Remediation — The Task Ahead,* University of Tennessee, Knoxville, TN, 1991.
197. Gardner, Frank, "Corrective Action Management Unit Rule," in *Corrective Action Training Course,* U.S. Environmental Protection Agency, Region 9, San Francisco, CA, 1993.
198. Godschalk, D. R., and Brower, D. J., "Mitigation Strategies and Integrated Emergency Management," in *Public Administration Review,* No. 45, Special Issue, 1985, pp. 64–71.
199. Charles, Michael T., and Kim, John Choon K., *Crisis Management — A Casebook,* Charles C Thomas, Springfield, IL, 1988.

200. Waller, R. A., and Covello, V. T., Eds., *Low-Probability/High-Consequence Risk Analysis: Issues, Methods, and Case Studies*, Plenum Press, New York, NY, 1984.
201. Kasperson, Roger E., and Pijawka, K. David, "Societal Response to Hazards and Major Hazard Events: Comparing Natural and Technological Hazards," in *Public Administration Review,* No. 45, Special Issue, 1985, pp. 7–18.
202. Petak, William J., "Emergency Management: A Challenge for Public Administration," in *Public Administration Review,* No. 45, Special Issue, 1985, pp. 3–7.
203. Zimmerman, R., "The Relationship of Emergency Management to Governmental Policies on Man-Made Technological Disaster," in *Public Administration Review,* No. 45, Special Issue, 1985, pp. 29–39.
204. Kelly, R. B., *Industrial Emergency Preparedness,* Van Nostrand, New York, NY, 1989.
205. Kelly, R. B., *Industrial Emergency Preparedness,* Van Nostrand, New York, NY, 1989.
206. U.S. Environmental Protection Agency, *Title III List of Lists: Consolidated List of Chemicals Subject to Reporting under the Emergency Planning and Community Right-to-Know Act,* U.S. EPA, Washington, D.C., 1991.
207. Environmental Management Services, *Title III Handbook: Emergency Planning and Community Right-to-Know,* DuPont, Wilmington, DE, 1987.
208. U.S. Environmental Protection Agency, *SARA Title III Fact Sheet: Emergency Planning and Community Right-to-Know,* U.S. EPA, Washington, D.C., 1990.
209. U.S. Environmental Protection Agency, *Toxics in the Community: 1988 National and Local Perspectives,* U.S. EPA, Washington, D.C., 1990.
210. Federal Register, "Federal Compliance with Right-to-Know Laws and Pollution Prevention Requirements," *50 FR 41981,* August 6, 1993.
211. Federal Register, "Risk Management Programs for Chemical Accidental Release Prevention," *58 FR 54190,* October 20, 1993.
212. Briese, Byron L., and Griffith, J. Gary, Eds., *Fire Protection Management for Hazardous Materials: An Industrial Guide,* Government Institutes Inc., Rockville, MD, 1991.
213. Baldwin, Dawn A., *Safety and Environmental Training — Using Compliance to Improve Your Company,* Van Nostrand Reinhold, New York, NY, 1992.
214. Cahill, Lawrence B., and Kane, Raymond W., Eds., *Environmental Audit — Sixth Edition,* Government Institutes Inc., Rockville, MD, 1992.
215. National Response Team, *Hazardous Materials Emergency Planning Guide,* NRT, Washington, D.C., 1987.
216. Gabor, T., and Griffith, T. K., "The Assessment of Community Vulnerability to Acute Hazardous Material Incidents," in *Journal of Hazardous Material,* Volume 3, Number 4, 1980, pp. 323–333.
217. Pijawka, K. David, Foote, Steve, and Soesilo, J. Andy, *Risk Assessment of Transporting Hazardous Materials, A Conceptual Framework and Approach for Risk Management and Planning,* Arizona State University, Tempe, AZ, 1984.
218. Pijawka, K. David, Foote, Steve, and Soesilo, J. Andy, "Risk Assessment of Transporting Hazardous Material: Route Analysis and Hazard Management," in *Transportation Research Record,* Number 1020, 1985, pp. 1–6.
219. Hall Jr., Ridgway M., *RCRA Hazardous Wastes Handbook — Ninth Edition,* Government Institutes Inc., Rockville, MD, 1991.
220. Lukaszewski, James E., and Serie, Terry L., "Public Consent Built on Credibility is the Goal," in *Waste Age,* February 1993, pp. 45–54.

221. Harrison, Bruce E., *Environmental Communication and Public Relations Handbook — Second Edition,* Government Institutes Inc., Rockville, MD, 1992.
222. Lukaszewski, James E., and Serie, Terry L., "Relationship Built on Understanding Core Values," in *Waste Age,* March 1993, pp. 83–94.
223. McDermitt, David, "The Ten Commandments of Community Relations," in *World Wastes,* Volume 36, Number 9, 1993, pp. 48–51.
224. U.S. Environmental Protection Agency, *Decision-Makers Guide to Solid Waste Management,* U.S. EPA, Washington, D.C., 1989.
225. U.S. Environmental Protection Agency, *Sites for Our Solid Waste — A Guidebook for Effective Public Involvement,* U.S. EPA, Washington, D.C., 1990.
226. U.S. Environmental Protection Agency, *Reducing Risk — Setting Priorities and Strategies for Environmental Protection,* U.S. EPA, Washington, D.C., 1990.
227. Burke, Thomas A., "Overview: Refining Hazardous Waste Site Policies," in Richard B. Gammage and Barry A. Berven, Eds., *Hazardous Waste Site Investigations — Toward Better Decisions,* Lewis Publishers, Boca Raton, FL, 1992.
228. U.S. Environmental Protection Agency, *RCRA Public Involvement Manual*, U.S. EPA, Washington, D.C., 1993.
229. Hartzog, Roger L., "New Federalism and State and Local Environmental Enforcement," in *Western States News,* Volume IV, Issue 1, 1993, pp. 1–9.
230. U.S. Environmental Protection Agency, *Local Government Implementation of Environmental Mandates — Five Case Studies, Final Report,* U.S. EPA, Washington, D.C., 1993.
231. U.S. Environmental Protection Agency, *Guidance for Implementing the FY 1994 State/EPA Enforcement Agreements Process,* U.S. EPA, Washington, D.C., 1993.
232. Fletcher, Jeff, "Environment is Top Concern of Local Leaders," in *Environmental Decision*, February 1990, p. 14.
233. U.S. Environmental Protection Agency, *A Preliminary Analysis of the Public Costs of Environmental Protection: 1981–2000,* U.S. EPA, Washington, D.C., 1990.
234. Steinzor, Rena, "Environmental Protection in the '90s: How Should Local Government React?" in *Environmental Decision*, February 1990, pp. 18–31.
235. Air/Water Pollution Report, *The Changing Face of Environmental Policy in the U.S. — Outlook and Trends,* Air/Water Pollution Report, Silver Spring, MD, 1994.
236. U.S. Environmental Protection Agency, *Enforcement Accomplishments Report FY 1992,* U.S. EPA, Washington, D.C., 1993.
237. Environment Reporter, "Congressional and Administrative Outlook for Environmental Issues in 1994," *Environment Reporter,* January 21, 1994, pp. 1660–1684.
238. Bullard, Robert, "Environmental Justice for All," in *National Wildlife EnviroAction,* November 1991, pp. 6–10.
239. Harding, Anna K., and Holdren Jr., George R., "Environmental Equity and the Environmental Professional," in *Environmental Science Technology,* Volume 27, Number 10, 1993, pp. 1990–1993.
240. Hazardous Waste News, "Clinton Signs Executive Order to Push Environmental Justice," *Hazardous Waste News,* February 15, 1994, p. 50.
241. Environment Reporter, "Environmental Regulation Does Not Lead to Weaker State Economies," *Environment Reporter,* October 1, 1993, pp. 1004–1005.

INDEX

A

B

C

O

P

S

U

V

W